FARRAR
STRAUS
GIROUX

Olives

Olives

The Life
and Lore
of a
Noble Fruit

Mort Rosenblum

North Point Press

Farrar, Straus & Giroux

NEW YORK

Designed by Abby Kagan

First edition, 1996

Grateful acknowledgment is made for permission to reprint the following: Excerpts from
An Orderly Man by Dick Bogarde with permission of Alfred A. Knopf. Excerpts from *An
Orderly Man* by Dick Bogarde reprinted in the United Kingdom with permission of the
Peters Fraser & Dunlop Group Ltd. Copyright © 1983 by Labofilms S.A. Excerpts from
Noé by Jean Giono with permission of Editions Gallimard. Copyright © 1961 by Editions
Gallimard. Excerpts from *The Mermaid Madonna* by Stratis Myrivilis, translated by
Abbott Rick with permission of Random House UK. Copyright © 1959 by Hutchinson
& Co. (Publishers) Ltd. Excerpts from *The Mediterranean Diet* by Nancy Harmon Jenkins
with permission of Bantam Books, a division of Bantam Doubleday Dell Publishing
Group, Inc. Copyright © 1994 by Nancy Harmon Jenkins. Excerpts from *Mediterranean Cooking* by
Paula Wolfert with permission of HarperCollins. Copyright © 1994 by Paula Wolfert. Excerpts from
Mediterranean Flavours by María José Sevilla with premission of Pavilion Books and David Higham
Associates. Copyright © 1995 by Pavilion Books. Excerpts from *From the Land of Figs and Olives* by
Habeeb Salloum and James Peters with permission of Interlink Books, an imprint of Interlink
Publishing Group. Copyright © 1995 by Habeeb Salloum and James Peters. Excerpts from *The
Food and Wine of Greece* by Diane Kochilas with permission of St. Martin's Press. Excerpts from
The Food and Wine of Greece by Diane Kochilas reprinted in the United Kingdom with permission
of the Doe Coover Agency. Copyright © 1990 by Diane Kochilas.

Library of Congress Cataloging-in-Publication Data
Rosenblum, Mort.
        Olives : life and lore of the noble fruit / Mort Rosenblum. — 1st
ed.
            p.   cm.
        ISBN 0-86547-503-2 (alk. paper)
        1. Olive—History.   2. Olive industry and trade—History.
I. Title.
SB367.R64   1996
641.3'463—dc20   96-20764
                                                                    CIP

Third printing, 1997

For Jeannette. And Louie Kay

Acknowledgments

Authors inevitably owe more thanks than they can possibly enumerate. Writing about olives, I discovered, carries this to the extreme. Generosity is a common trait among olive people. Many of those who trailed me through their groves, took me into their homes, fed me their oil, and remembered bits of forgotten lore are mentioned in the text. To single out a few would be to leave out others. Nonetheless, I am deeply grateful to the Romanas, Bosquets, Martins, Beuguehos, and Casanovas of Ampus.

Particular thanks go to those who may not appear in these pages: Dev Kernan, master builder and faithful friend, gave me that first copy of *Le Livre de l'Olivier* which set me on this path. Friends like Gretchen Hoff, Chris Dickey, Olivia Snaije, Melissa Eddy, Hugh Mulligan, Paul Theroux, Phil Cousineau, Gregory McNamee, Barbara Porter, Lito Tejada-Flores, Constance Kyriazakou, Debra Williams-Gualandi, and Dennis Redmont, among so many others, fed me regularly with oliviana. Willis Barnstone and Nancy Harmon Jenkins reappeared from the past to lend a hand.

My agent, Geri Thoma, Joan of Arc with a sense of humor, caused the book to happen. Mary Scarvalone's pencil drawings made it beautiful. Becky Gallagher shepherded the detail work. And my editor, John Glusman, understood that manuscripts need more pruning than olive trees.

At Wild Olives, Monsieur and Madame Beugueho watched over the place when I was off picking other people's trees. And above all, Jeannette Hermann kept me on track during the tough times when I was ready to chuck it all and go back to butter.

Contents

1	Olives	5
2	Wild Olives	23
3	Mount of Olives	49
4	Olive Heaven	75
5	Olives of Wrath	97
6	Top-of-the-Line Tuscan	111
7	Cosa Nostra	137
8	Running of the Olives	163
9	Socialized Olives	179
10	Marrakesh	203
11	The Lesbos Groves	217
12	The Big Kalamata	233
13	The Olive Branch of War	257
14	The Noble Ingredient	267
15	California Olive Rush	285
Epilogue	Another Season	313

Olives

Kalamata Olives

". . . I like them all, but especially the
olive. For what it symbolizes, first of all—
peace with its leaves and joy with its golden oil."
—*Aldous Huxley*

"The olive tree is surely the richest gift
of Heaven."
—*Thomas Jefferson*

1

Olives

An olive, to many, is no more than a humble lump at the bottom of a
martini. Yet a closer look reveals a portrait in miniature of the richest parts
of our world. Olives have oiled the wheels of civilization since Jericho built
walls and ancient Greece was the morning news. From the first Egyptians,
they have symbolized everything happy and holy in the Mediterranean.
But it is simpler than all that. Next time the sun is bright and the tomatoes
are ripe, take a hunk of bread, sprinkle it with fresh thyme, and think
about where to dunk it. I rest my case.

Unforeseen circumstances pulled me into olivedom, and my conver-
sion was swift and surprising. Olives had meant nothing to me. Olive oil
was an overpriced, hyperpungent, pretentiously packaged fluid you dribbled
into pasta sauce when in an ebullient mood. I figured green olives grew on
one kind of tree and black on another.

Then, in 1986, I bought a pile of rocks and five acres of Provence jungle. Along with bramble thickets, yellow-flowering broom, old oaks, and charred beams of a ruined farmhouse came two hundred half-dead, over-grown olive trees that were already old when the Sun King ruled France. Ancient Greeks planted the neighborhood's first groves, and the roots of some trees down the road were rumored to have been growing before Christ was born.

I could spend whole afternoons looking at old man Romana's trees, the lords of our mountainside. Nature rarely got so imaginative, or beau-tiful. Timeless olives stretched out of sight down a steep slope, each trunk craggier and more permanent-looking than the old stone walls between the even rows. Every tree was shaped for style and easy picking: three main branches were pruned each spring to sprout the graceful new wood that produces fruit. In sunlight, the leathery gray-green leaves flashed silver. At dawn, they glistened with dew. At dusk, the last pink light tinted them the pastel shades that Renoir captured in his old orchard not far away at Cagnes-sur-Mer.

Romana himself seemed as timeless as his trees, bent from eight dec-ades of hard life. Olives had a way of attracting such strong sorts, who these days were otherwise found only in scratchy black-and-white films and leatherbound novels.

On crisp evenings, we cooked lamb and fish in the fireplace over olive-wood coals. The twisted logs threw off bright yellow flames, with no smoke, until they burned down to a nightlong bed of coals that scented anything on the grill. In the morning, nothing but a fine white ash re-mained. We used only hollow trunks, burls, and odd pieces. Olive wood is too handsome to squander; the whorls of its grain polish to works of art. At Christmas, a craftsman friend who came to visit carved handles for silverware from old branches we had cut away.

One morning I brought my first meager crop to the sixteenth-century mill at Tourtour and watched two Pagnol characters operate the water-driven stone wheel and oak-beam press. When the cloudy golden liquid oozed into a clay urn, I dipped in a bread crust to check the flavor. I was ready to write psalms.

Until then, I'd been absorbed with pruning saws and sheep droppings. With my new oil in the pantry and my trees resting for the winter, I stepped back to see a larger picture. Olives, like grapes, are essential to any life worth living. But you can't see by the light of burning wine, or massage a friend's temples in grape juice, or heat a house with vines.

Olives have permeated every Mediterranean culture from prehistory to last week. Aristotle philosophized about them, and Leonardo invented a modern way to press them. Egyptian pharaohs were sealed into pyramids with golden carvings of olives. Greeks used so much oil to lubricate their athletes that they devised a curved blade, the *strigil*, to scrape it off. The earliest Olympic flame was a burning olive bough. Rome had a separate stock market and merchant marine for oil. And conquering gladiators, like the Roman emperors, were honored with olives.

Olives were domesticated before anyone devised words to record the fact. They were around when the Bible was a rough draft scrawled on papyrus. Later wisdom was written and read by the clear gleam of olive-oil lamps. Long before it was food, olive oil beautified the body, cured ills, and fed the soul. For a time in Greece, only virgins and young men sworn to chastity were allowed to harvest the trees. When Odysseus finally came home, he collapsed into the marriage bed he had made Penelope from a massive olive trunk.

Saul, the first King of Israel, was crowned by rubbing oil into his forehead. In Hebrew, the root word for "messiah" comes from "unguent"; if one ever appears, he—possibly she—will be slathered in oil. The doors of Solomon's temple, and its huge four-faced winged cherubs, were carved in olive wood.

When the Jews declared the miracle of Hanukkah because their eternal flame flickered eight days on empty, their fuel was olive oil. The boiling oil that Galileans poured on Roman legions to repel their sieges was pressed from olives. It is a safe bet that the Three Wise Men came to the holy manger with cold-pressed virgin in their luggage. Today, still, a Hebrew idiom for a good man is "pure olive oil."

To Christians, the olive was equally sacred. Clovis, first King of the Franks, was crowned only because a dove from heaven appeared in the nick of time, reportedly, with a flask of oil for the holy rites. That was in 481; the pear-shaped vial, kept in the Basilica at Saint-Rémy, was used to similarly oil thirty-four more French monarchs. The Bible's two testaments refer to olive oil 140 times and to the olive tree nearly one hundred.

The Prophet Muhammad likened the holy light of Allah's being to the sparkling radiance of burning oil from "the Blessed Tree, neither of the East nor the West." Islam's oldest university, in Tunisia, is named al-Zitouna: The Olive Tree.

To Muslims, as to Christians and Jews, the olive means wisdom, fertility, and peace. But for thousands of years it has also symbolized conflict. Even now, olives are at the heart of Holy Land politics. Both sides plant them in the West Bank to mark possession. When Palestinians arise, Israelis riposte with terrible revenge. If a single stone flies from behind an olive tree, the bulldozers come. Ancient groves have gone under the blade, emasculating families for generations.

People who live among olive trees tell you their air is pure and their lives are full. They expect miracles as a matter of course. On her 121st birthday, Jeanne Calment of Arles, France, had a simple answer when asked how she survived to be the world's oldest-living person: olive oil. It appears in nearly every meal she eats and each day she rubs it into her skin. "I have only one wrinkle," she said, "and I am sitting on it."

For those who venerate them, olive trees are magical. Old trunks appear supernaturally sculpted over the centuries, twisted into shapes of wizened faces or runes of mystical meaning. To flourish, olives need the hand of man. Left untended, roots throw out shoots which suck vitality from the main trunks. You must decide what to sacrifice, cutting away new growth to save the mature core, balancing old and new. To revive a long-forgotten tree, you must whack away to within an inch of its life. The old Provençals gave the olive its own watchword: Make me poor and I'll make you rich.

There is something primal and wonderful about watching olives ripen. They begin changing color in October, first to a dappled light mauve and then through a spectrum of reds and purples. By December, they are black. The Tuscans have already picked by then, preferring a green, sharper oil. We wait until it is nearly Christmas. The olives are a satiny blue-black, fully mature, and bursting with sweet yellow oil.

One afternoon, dribbling this oil onto cucumbers in the shade of my favorite tree, Emiliano, I felt a strange kinship. This old character had survived a careless bulldozer driver's blade. A rock wall had toppled onto its trunk, nearly uprooting it. Now it was thick with fruit. Nearby, Julio was bright green with young buds, the result of clearing the competing brush and turning the earth. The air was heady with the scent of oleander, rosemary, lavender. In neighboring groves I heard the comforting growl of mototillers, chain saws, and weed whackers. With a little help from the industrial age, men with faces as weathered as their trees were doing what olive growers had done a millennium earlier. This was a subject to pursue.

Olives appeared in unlikely places. One friend, half Greek, told me about Soula. In prewar Athens, fathers gave their pubescent sons a handful of drachma, a condom, and the address of a good-hearted prostitute. My friend's father went to see Soula. Having no diaphragm, she relied on an alternative protection from pregnancy: a fat Kalamata olive.

Unexpected passions caught me off guard. I mentioned olives to James Lemoyne, a Miami-dwelling bon vivant with Latin American tastes, and he sat bolt upright. "I eat thirty to forty a day," he declaimed. "When I ate my first olive, I knew instinctively I was tasting the heart of the world. When I first tasted extra-virgin oil, I drank it down straight from the bottle. It felt like the very blood of the warm, rich earth."

My friend Chris Dickey, a Middle East and Mediterranean specialist for *Newsweek*, turned out to be a closet olive loony. The subject began to fascinate him years earlier in Andalusia; a priest could not explain why despondent youths preferred to end their lives by hanging from an olive tree. That, he knew, was how Marcel Pagnol's Ugolin exits the scene in *Manon des Sources*.

The Christopher Dickey Theory of Olive Politics neatly explains why much of the world took the shape that it did. He traced the olive's civilizing—and destabilizing—influence around the Mediterranean rim. The Middle East oil boom began not in 1973 but rather a thousand years before Christ, when Philistines controlled exports. From Phoenicians to Venetians, the oil trade was a key to power. Frederick II, the last great Hohenstaufen and Holy Roman Emperor of the thirteenth century, was a Teutonic tyrant born among the olives in Italy. He conquered Sicily and, in the Crusades, crowned himself King of Jerusalem.

For Dickey, the Olive Line explains why Arabs have never managed to make peace among themselves. Beyond the Jordan River, the desert starts, and Bedouin mentalities shift like the sands. In Palestine and the Levant, it is different. "Without our olives, we feel like a paper in the wind," a merchant in Jenin told him once. "The olive tree, it means the land. If you live twenty, thirty years in a place, you feel that in your breast, in your body. We take our own shape from our trees."

Your basic olive is an *Olea europaea*, a member of the only family in a small order of flowering plants called Oleales. The jasmine and lilac are blood relatives. So are ashes, hardwood trees of decidedly different haunts and habits. Naturalists can only guess at the olive's origins. Sir David Attenborough's account in *The First Eden* is as likely as any. Millions of years ago, the Mediterranean disappeared. Land shifts closed off Gibraltar, and the sea evaporated. Then, mysteriously, the straits burst open again and water came flooding back. Among the foliage on the islands and shores of the eastern Mediterranean wild olives appeared. Who knows when. Fossilized leaves dating from 37,000 years B.C. were found on the Aegean island of Santorini. Those scrubby oleasters still abound.

Around 6000 B.C., perhaps later, farmers in Asia Minor discovered that wild olive shoots could be grafted, replanted, and domesticated. Greek islanders also tamed the tree. Egyptians revered the olive, and one theory places its first roots in the Nile Delta, where it used to flourish. No one is certain. A few old legends trace the olive tree back to Adam.

But if science does not pinpoint the origin of the olive, mythology has its own explanation. When Zeus sought a deity to rule over Attica where the Acropolis would stand, he devised a contest. The god who gave mankind its most valuable gift would win. Poseidon struck his trident on a rock. A rearing horse emerged, capable of carrying people long distances, of winning wars, of hauling heavy loads. Athena produced an olive tree.

There is another version. Hercules (Herakles), son of Zeus and symbol of everything Mediterranean, thrust his staff into the bare ground. From it, olive leaves grew. "The upright Aitolian judge of the Greeks, obeying Herakles' ancient laws, loops the gray glory of the olive over the hero's brow and locks," Pindar wrote five centuries before Christ. "Long ago Herakles carried the silver olive tree from the shadowy spring of the Danube to make it the handsomest symbol of the Olympian games."

Willis Barnstone, who translated that piece of Pindar, is a friend who settled in to teach at Indiana University after half a lifetime around olives in Greece, Spain, and North Africa. He offered a few lines of his own on their origins: "God, needing a word to speak the cosmos into being, first created the alphabet. In those days he worked in Hebrew. Liking the shapes of those letters, he turned them into olive trees. Today, if you wish to read the mystery of creation, look deep at the calligraphy of an olive branch."

As time went on, olives were crucial to civilizations spreading from the Near East and around Crete. They were the currency of the Inner Sea, a cultural inspiration for its great empires: Assyrian, Egyptian, Greek, Persian, Roman. Among the most breathtaking relics of antiquity are olive *objets*: Grecian jars, Minoan frescoes, Egyptian bas-reliefs, Roman carved silver vases, Carthaginian mosaics. Motifs show harvesters in their trees, wielding thin rods, just as olives are often collected today.

During the Peloponnesian Wars, the olive trees that were a symbol of peace reshaped the style of battle. Towns grew behind impregnable walls. Attacking armies could lay siege and demoralize entrapped populations only by destroying their source of food. But, historian John Keegan notes, they were stymied by the olive groves. Scorched trees threw out fresh buds the following season. That meant digging out trees by the roots. Spartan invaders were warriors, not farmers, and refused to spend months with a shovel. And defenders, far from demoralized, burst forth in fury to protect their beloved olives. In the end, Attica fell only when its olives were finally devastated.

Greek colonizers brought the olive to Sicily between the eighth and fifth centuries B.C., using wealth from oil and wine to build the great city of Syracuse. They took olives across the turbulent strait between the monsters Scylla and Charybdis to the Italian mainland. Developing trade routes, Greeks and Phoenicians took the olive west to France, on to Spain and to Tunisia, on opposite shores of the Mediterranean.

The Latin *olea* took root from the Greek *elea*. The Romans planted huge groves in North Africa and France. In Spain, they hardly had to bother. When Caesar's legions assaulted Seville, their horses could barely make it through the olive trees ringing the city. But Moorish invaders brought so many more seedlings to Spain that the Spanish words for olive and oil—*aceituna* and *aceite*—derive from Arabic.

By the tenth century, groves fringed the Mediterranean across southern Europe and northern Africa, and they covered its islands. Spanish missionaries brought trees to the New World in the 1500s, behind the *conquistadores.* Then Italian immigrants carried the olive to South America, Australia, and southern Africa.

Today, there are about 800 million olive trees in the world. China has 20 million, four times as many as France. Small stands grow in Angola, in darkest Africa. They are found on six continents, but 90 percent of them fringe the Mediterranean. Italy springs to mind at the mention of olive oil, but Spain has more trees—a lot more, if you deduct the fictitious groves Italians report to collect European Union subsidies—and oil labeled as Italian is often from Spain, Greece, Turkey, or Tunisia.

Olives account for at least 200 million workdays a year, and perhaps 7 million families grow them for harvest. Many are weekend farmers who tend their great-grandfathers' trees for the love of it. Others might own a hundred thousand trees. Worldwide, the oil and olive business adds up to a turnover that approaches $10 billion.

Each year, the world consumes nearly 2 million metric tons of olive oil. In Greece, every man, woman, and infant averages five gallons. Elsewhere, the numbers are climbing fast, as more people acquire the taste; in America, the recent annual increase has averaged 12 percent.

With each new survey doctors are more enthusiastic about the nutritional value of olive oil. Monounsaturated, it drives off the bad cholesterol without reducing the good. Unlike animal fat, it does not linger in the body as a cancer risk; evidence shows it wards off certain cancers. It aids digestion, helps children grow, and retards aging in the bones, joints, and skin. Although experts quibble, all say that olives are good for health. The only question is how good.

Encouraged by the new popularity of olive oil, small independents joined the multinationals and the Mafia to expand the market. Designer bottles sprouted everywhere. Suddenly a new breed of connoisseurs could shop for new nectars to keep in reserve, waiting for the perfect tomato.

Used alone, the word "olive" is hardly more specific than "grape," and olive oil is as nuanced as wine. The *Olea europaea* comes in at least seven hundred cultivated varieties, or cultivars, and each produces a different sort of fruit. Tough little Cypriot strains cling to rocky hills like petrified wood with leather leaves. Our delicate caillet roux, *la plante de Figanière*, dips like a willow. Its olives hang low in thick bunches, red as cherries, until ready for picking.

The best oil producers blend varieties to capture a distinctive flavor, but even that is no guarantee. It all depends on the rains, the pests, the time of harvest, the pressing process, and how the oil is stored. Unlike wine, oil does not improve with age. It tends to be sharp when fresh. Within a few months it has pleasurably mellowed. After a year, many oils edge toward rancid. A few, like the Spanish picual, last a few years longer, but only when absolutely pure and kept away from light and heat.

Once, in Spain, I asked an olive expert which was the best oil. "What is the best cheese?" he replied. It is all a matter of taste, and no broad categories apply. "Italian" oil ranges from syrupy yellow southern oils to thin green Tuscan *crus* with a "peppery" (that can mean bitter) afterbite that may last through a long siesta. In an hour's drive, you can go from the sweet, buttery oils of Liguria, past the green and fruity Luccas, to the sharp elixirs of high Chianti. Fanatics might keep a dozen oils in their kitchens, from Cretan to Californian.

A basic set of standards guides the consumer. Extra-virgin oil means that the amount of free fatty acids—mostly oleic acid—is below 1 percent. Also, the organoleptic properties—taste, aroma, feel on the tongue—must rate high. Virgin oil, rarely found on the market, can have up to 2 percent acidity. Both are freshly squeezed by one of several processes known as "first-press" or "cold-press."

Plain olive oil, often marked "pure," is refined inferior stuff best kept for frying. It is "rectified" with steam and chemicals and then mixed with better oil for a little flavor and aroma. Pomace oil comes from the first-press leavings, refined to bring it below the 3.5 percent acidity level that designates lamp oil. These rules were established by the International Olive Oil Council, a Madrid-based agency backed by the United Nations. They apply only if what is on the label is in the bottle.

Like grapes, there are olives for pressing and for eating. Unlike grapes, you cannot put fresh-picked olives on the table unless you are troubled with sadism; they are excruciatingly bitter. It can take months to prepare an olive properly, and therein lies its richness. The industrial method produces those canned "black ripe" globules, graded from huge to humongous. Doused in caustic soda for a quick cure, and bubbled with an iron compound for color, little remains of taste or texture.

Every culture cures olives differently. The Moroccans alone have a hundred ways to do it. For an idea of the range, drop by any weekly market in southern France. To taste chewy black olives in *herbes de Provence*, or cracked green picholines in chilies and lemon, or tart little Niçoise, is to understand why the blessed tree has turned so many heads since history began.

In the sixth century B.C., the poetess Sappho was already inspired by the voluptuous little fruit at her academy for young women on the island of Lesbos. Two and a half millennia later, Lord Byron exalted Grecian groves, not far away. Olives have inspired literary flights from Azerbaijan to Andalusia. "Black angels were flying through the sunset wind," Federico García Lorca wrote early this century: "Angels with long braids and hearts of olive oil."

Most early literature touched on the technical. "The olive has spread even across the Alps and into the heart of Gaul and Spain," Pliny the Elder recorded in *Natural History*, in A.D. 50, summing up data that had emerged since the Philistines, two thousand years before him. Pliny quoted a report from 581 B.C. which said there were then no olives in Italy, Spain, or Africa. A century later, Theophrastus, the Greek, observed—wrongly —that the olive grew no farther than forty miles from the sea.

Pliny was a devotee of sensible Mediterranean behavior: sip the wine and splash the oil. "Age imparts an unpleasant taste to oil, which is not the case with wine, and after a year it is old," he wrote. "Nature shows forethought in this, if one chooses to interpret it this way, since it is not necessary to use up wine which is produced for getting merry; indeed, the pleasant overripeness that comes with maturity encourages us to keep it. She did not, however, wish us to be niggardly with oil, and has made its use widespread, even among the masses, because of the need to use it up quickly."

The ancients liked their oil green, from olives picked just after the grape harvest. "The riper the berry the more greasy and less pleasant is the flavor of the oil," Pliny wrote. "The best time for gathering olives, striking a balance between quality and quantity, is when the berries begin to turn dark—locals call these *druppae*, the Greeks, *drypetides*." In much of the world, this guideline is still the fashion.

Cato's instructions, as relayed by Pliny, might have been written today. He expounded on planting, pruning, pressing, down to a hundred uses for the black vegetable water that separates from pure oil. The trees those old Romans left behind have inspired every generation since.

In Spain, under the Romans, Lucius Columella called the olive the Queen of Trees, an apt title not only for its practical uses but also for the symbols it conveys in literature. Ever since a dove brought an olive sprig back to Noah's Ark, it has meant peace. It stands for strength: Hercules' staff was an olive trunk. And sacrifice: Christ was seized on the Mount of Olives and nailed to an olive-wood cross. Olympic victors wore olive wreaths and were rewarded with oil from sacred trees; this symbolized victory and wealth. It signifies the transfer of power. Kings and emperors were anointed with oil. And, from Athena, it means fertility.

Today, every olive society has its classic reference works. Deep within the French Bibliothéque Nationale, for instance, is A. Coutance's *L'Olivier*, published in 1877. It is thick with religious symbolism and ancient anecdotes. Coutance recounts the founding of Athens and adds the sequel. When Xerxes attacked from Persia, he destroyed the Acropolis and burned Athena's tree. The Greeks returned to find only rubble on their holy hilltop. But rain had washed ashes from the sacred roots, which sprouted again. This, Coutance said, was yet another symbol: resurrection. Only olive trees do that.

At the Carli Fratelli olive museum at Imperia, between Genoa and Nice, cracked leather tomes from the 1700s nestle against leaflets on modern pest control. An old bibliography on oliviana covers 120 letter-size pages, in cramped, small type, with a list of works scattered around the world. And that is only up to 1943.

Cosimo Moschettini, in his 1794 masterwork, *Della Coltivazione degli Ulivi e della Manifattura dell'Olio*, probes with style and grace the olive's most intimate fertility secrets. He gives Virgil, Pliny, and Cato their untranslated say. Raffaello Pecori, in *La Cultura dell'Olivo in Italia* from 1891, covered the world olive economy: "Australia and California have embarked with faith and courage in the culture of this plant. The United States government is particularly anxious to support it."

Modern writers tend to be lighter-hearted or more romantic about the subject. The grand French historian Fernand Braudel defined the Mediterranean as "where olives grow," but a whimsical Ford Madox Ford was more specific: "Somewhere between Vienne and Valence, below Lyons on the Rhône, the sun is shining, and south of Valence [the] Provincia Romana, the Roman Province, lies beneath the sun. There is no more any evil, for there the apple will not flourish and the brussels sprout will not grow at all."

Provence is storybook olive country, and its authors can get a little carried away. Jean Giono, a late chronicler of Provençal life, wrote that the mere whiff of olive oil exempts one from reading Homer. In *Noé*, he exulted in the joy of climbing high in his trees in a freezing December to imagine strange Odyssean landfalls from a wine-dark sea and to ponder the meaning of life.

There, I learn a serious lesson in greed [he wrote]. By nature, I'm not greedy. There are, on the contrary, a hundred thousand reasons why I am not. But I could be rent by the cold (I am rent by the cold), and yet not come down from my trees; I would not stop picking. My hands stick like glue to my olives. Should God suddenly close the world like a book and say: It is over; should the trumpet call the dead, I would appear for judgment caressing olives in my pocket; and, if I had no pocket, I'd caress olives in my hands; if I had no hands, I'd caress them with my bones, and if I had no bones, I'd find a way to caress them still, if only in spirit. If that is not greed, what is?

And the worst of it. If you say to me, give me a sack of olives, I would give two sacks. But if you say, "Let me climb that tree, let me pick that fruit, let me take it in my hands in your place," I would resist to the last Judgment, I would resist God Himself, and I am sure of finding, to resist Him, the force to triumph, to the point of becoming myself capable of a miracle. What a remarkable food!

Aldous Huxley's jewel of an essay, "The Olive Tree," was written in Provence in 1936. Developing the thesis that all cultures at one time worshipped trees, he wrote:

> Solidified, a great fountain of life rises in the trunk, spreads in the branches, scatters in a spray of leaves and flowers and fruits. With a slow, silent ferocity the roots go burrowing down into the earth. Tender, yet irresistible, life battles with the unliving stones and has the mastery. Half hidden in the darkness, half displayed in the air of heaven, the tree stands there, magnificent, a manifest god . . . I like them all, but especially the olive. For what it symbolizes, first of all—peace with its leaves and joy with its golden oil.

It matters little that what the olive stood for frequently meant only "the peace of victory, the peace which is too often only the tranquillity of exhaustion or complete annihilation," he wrote. What we remember is that Roman conquerors faced their ovations in a crown of olive leaves. So did the Greeks.

Huxley thought too little attention was paid to the role olives played among the ancient Hebrews, who attached religious, social, and sensuous significance to glistening fat. It anointed their kings. But their thin pastures could not support cows to produce butter and suet. "Only the sheep and olive remained as sources of that physiologically necessary and therefore delicious fat in which the Hebrew soul took such delight."

The olive, Huxley concludes, is the painter's tree. "Under a polished sky the olives state their aesthetic case without the qualifications of mist, of shifting lights, of atmospheric perspective, which give to the English landscape their subtle and melancholy beauty . . . [The olive] does not need to be transposed into another key, and it can be rendered completely in terms of pigment that are as old as the art of painting."

But olives humbled the painters, and they terrified some. Vincent van Gogh traveled south and wrote to his brother in 1889: "Ah, my dear Theo, if you could see the olives at this moment . . . The old silver foliage and the silver-green against the blue. And the orange-hued turned earth. They are totally different from what one thinks in the north . . . The murmur of an olive grove has something very intimate, immensely old. It is too beautiful for me to try to conceive of it or dare to paint it."

Van Gogh painted eighteen canvases of noble olives, in blues and deep greens. Cézanne did them from a distance, garnish to his beloved Mont Sainte-Victoire. But Huxley named Renoir, whose paintings at Cagnes gave the gray trees shadows of cadmium green and a suffused glow of pink.

Renoir, in fact, only rarely painted his olives. Like Van Gogh, he found them overpowering. The slightest breeze caught their glimmering leaves, constantly changing their shape and color. "Look at the light on the olives," the aging Impressionist wrote one afternoon. "It sparkles like diamonds. It is pink, it is blue, and the sky that plays across them is enough to drive you mad."

Instead, at Cagnes Renoir stayed indoors in his spare, narrow wheelchair, brush held in an arthritic hand as gnarled as an olive whorl. His trees were for inspiration and contemplation.

Renoir visited Cagnes at the turn of the century and found impending calamity. A merchant of Nice had acquired an old grove at the edge of town and was about to sell its majestic trees for charcoal. Renoir bought it. Having left his beloved Seine, he found a winter place where the sun would warm his aching joints. Mostly, however, he wanted to save the olives. They're still there, badly in need of a good pruning, but healthy, living monuments to themselves.

Other groves fared badly. "During the last few years there has been a steady destruction of olive orchards," Huxley wrote. "Magnificent old trees are being cut, their wood sold for firing and the land they occupied planted with vines. Fifty years from now, it may be, the olive tree will almost have disappeared from southern France, and Provence will wear another aspect."

Twenty years after Huxley wrote those lines, Lawrence Durrell moved to southern France and found the olive trees in serious trouble. Looking for a place to nest, Durrell settled in a stone house by a village in the Drôme. He had a hundred old trees. Nearly all their trunks were dead, split open in the winter that froze Provence. But Durrell arrived in 1957, a year after the freeze, and these were olive trees. Already, healthy green shoots sprouted from the roots. If tended, they would produce olives within four years. Within a decade they would be respectable trees, ready for another five centuries.

Durrell's taste for olives developed early, when his mother brought the family to Corfu for a long season. He spent most of his life in the Mediterranean landscapes, which kept drawing him back. His characters nap against olive boles. In *Panic Spring*, olives come alive, "moaning and dragging at their roots."

His shabby romantic in *Justine* brings warmth to a cold day in Alexandria with a small can of olives, spotted in a grocer's window, and bought with his last few coins: ". . . sitting down at a marble table in that gruesome light I began to eat Italy, its dark scorched flesh, hand-modelled spring soil, dedicated vines." Justine arrived, and he gave her an olive.

Late in life, in *Spirit of Place*, Durrell describes his pilgrimage to Corfu to bring his own green oil to Saint Arsenius in a rock niche on the rugged coast: "The silver olives slide breathlessly down in groves below one as if to plunge into it and to swim for dear life out into the blue . . . Yes, it was in this landscape that I learned many important lessons: the kind you cannot put into words. Even the driver had fallen silent now as we traversed these long, silent groves of ancient olive trees."

And, in *Prospero's Cell*, he explains why he was moved to such trouble: "The whole Mediterranean, the sculpture, the palms, the gold beads, the bearded heroes, the wine, the ideas, the ships, the moonlight, the winged gorgons, the bronze men, the philosophers—all of it seems to rise in the sour, pungent taste of these black olives between the teeth. A taste older than meat, older than wine. A taste as old as cold water."

Americans, by and large, have a different view of the olive. This disturbs Willis Barnstone, whose forty books touch regularly on the Mediterranean soul. Olives are often on his mind. "If there are four elements in the world—earth, water, fire, air—then the olive has to be the fifth," he said. Willis has a liquid laugh that lightens everything he says. One never knows when he is toying with hyperbole. On this subject, he is not.

"People who know olive trees revere them, like angels that spring from the earth," he said. "They live off them in the best way. The olive is to the Mediterranean what the camel is to the desert. Every tree is an individual, anarchic, a struggling survivor. To Americans, the olive is just a crop, and it is grown that way. Only Americans could name a silly cartoon character after the olive. No reverence."

It is not just any character. Olive Oyl, created in 1919 for the Thimble Theater, is the *grande dame* of cartoons. She is older than Minnie Mouse. Her brother, Castor Oyl, has slipped from memory, but Olive is as well known as Popeye. E. C. Segar, her creator, died in 1938, and I could not find anyone who knew why he gave her that name. But Willis was right. The Spanish cannot bear to translate Olive Oyl. They call her Rosario.

Still, Americans earned their place in olivedom. By some accounts, this traces back to an evening in 1870 when a California bartender named Julio Richelieu wanted to satisfy a miner who demanded something unusual for his gold nugget. He mixed gin with an aromatized white wine and ceremoniously plopped in an olive. His bar was in Martínez, so he called it the Martínez cocktail.

Or perhaps the honors go to a bartender at the Knickerbocker Hotel, who made a concoction in 1910 for John D. Rockefeller, with a twist of lemon and an olive. His name was Martini di Arona di Taggia; fortunately, only the first part stuck.

By 1943, everyone knew about the drink. When Franklin D. Roosevelt met with Winston Churchill and Joseph Stalin in Tehran, he offered each a "dirty martini": two parts gin, one part vermouth, and a dose of olive brine. Roosevelt also mixed one for Stalin's foreign minister, Vyacheslav Molotov, who apparently was not impressed. He devised his own cocktail, of greater potency.

Like the olive itself, the martini's true origins are lost in history. But so what? After eight thousand years of healing, illuminating, nourishing, enriching, and inspiring, the olive's place at the bottom of a glass is no bit part. Without its ennobling presence, a martini is no more than gin and vermouth.

Caillet Roux

Tapenade

Nothing is more basic to French olive people than *tapenade*, the simple and delicious Provençal caviar eaten with toasted bread. It can be flavored with garlic, bay, thyme, mustard, and even rum or Cognac. Around Wild Olives, it is done the old way.

Pitted black olives, salt-cured but unflavored

Anchovy fillets, washed free of salt or oil-cured

Capers

Mild extra-virgin olive oil

Salt and pepper to taste

Mash the olives, anchovies, and capers with a mortar and pestle, marble by tradition. The rule of thumb is a little more olives than capers and anchovies, but adjust according to taste. Add very little salt and pepper. Drizzle in the oil until creamy.

Aïoli

Aïoli is a Provence staff of life, an olive-oil mayonnaise served with poached cod, snails, boiled eggs, beets, cauliflower, potatoes, carrots, artichokes, and chickpeas. Patricia Wells, a food lover who has mastered the French way, does it like this:

6 large fresh garlic cloves, with the veins removed

2 large egg yolks, at room temperature

½ teaspoon salt

1 cup extra-virgin olive oil, the fresher the better

Warm the mortar with boiling water, then dry. Mash the garlic and salt to a paste. Add one egg yolk. Stir, pressing slowly and evenly, in a circular motion. Add second yolk. Work in the oil, drop by drop, until the mixture thickens. Whisk in the remaining oil in a slow stream. When the *aïoli* holds the pestle upright, it is ready.

Makes about 1 cup.

"Fai mi paure, ti fairai riche."
Make me poor, I'll make you rich.

"Espeio mi, ti vestirai."
Strip me, I'll dress you.

"Grate moun ped, ti graisse toun bec."
Scratch my foot, I'll grease your chin.
—Provençal proverbs; the
olive tree is speaking

2

Wild Olives

Back in the 1980s I wouldn't have known an olive branch if Noah's dove had flown up with one in its beak. I bought my little ruin in a back corner of high Provence for the beauty and peace, not the vegetation. It took a day's hacking just to reach the collapsed house. What with thick climbing vines and dense undergrowth thoroughly mined with spiky brambles, I had to take it on faith that olives lurked somewhere out there in the jungle.

With chain saw and machete I freed them one by one, energized by the rush that firemen must feel when digging survivors out from a collapsed building. Every victim was in a perilous state: starved, strangled, fungus-ridden, and feral. Some trees still bore fruit and needed only reshaping. Others would need drastic amputation and years of patient care. I named the place Wild Olives and set to work.

Quickly I passed though the romantic stage and adjusted to hard labor. Perusing my mounting pile of books, I realized that this was a common rite of passage.

In *An Orderly Man*, Dirk Bogarde describes how he tired of acting and settled in a home where he could write in tranquillity. He bought an old stone house high up above Grasse, surrounded by four hundred giant olives. His first surprise was that the trees were dying, drowning in water-logged soil. A wall had cracked in the Cannes city reservoir farther up the hill. All it took was a fortune to pump the land and an act of God to convince the Cannes town fathers to patch the leaks.

Once the land was drained to its natural state, Bogarde found the ground too rocky for farming, so he decided to direct his agricultural urges toward his trees.

They'd been there for centuries bearing their crops and, apparently, flourishing [he wrote]. I should make them do the same for me. Exactly how, at that moment, I did not know. I had some hazy idea that it was a sort of Christ-like business: disciples, and loaves and fishes. Ageless somehow. Figures seemed to drift before me in flowing homespun, baskets on arm, bare feet treading beneath the venerable trees, culling the ripe fruits with slender Veronese hands. A Biblical style of life which had, in fact, existed long before that Book had even been written. Timelessness. Calm. Absolute rubbish: of course.

Had I been told then and there, that morning on the terrace, with my father swigging his French beer and swatting wasps, that my Biblical image would be cruelly shattered by the inescapable fact that the harvest took place in the four bitterest months of the year, December to March, that I'd spend my time, not wandering in homespun, bare foot, from burdened tree to burdened tree with basket on my arm, but crouched on my knees in an anorak, sodden, frozen, fingers white with ice, gathering up the blasted little fruits, one by one, mice-nibbled, worm-ridden, in a howling mistral, I might have easily caved in, moved to an hotel and pressed the button for Room Service for the rest of my life.

Well, yes. Then again, I learned during my first few years, it makes a fine Christmas tradition. You simply find a few friends who have forgotten their Tom Sawyer and invite them down for the holidays. At midday, half dead and hating you, they sit down to steaming pasta in a sauce so rich in oil that yellow puddles settle on top. That gets them going until dark, when it's time for Mediterranean bass grilled on olive-wood coals. Beat that with room service.

Nothing brings you closer to your trees than that Christmas communion. Crawling under twisted branches to reach the olive bunches, you notice the nuances. Ernesto responded magnificently to spring pruning, but Doorman is turning black from a dreaded fungus called *fumagine*. Darth looks a little less menacing with fresh leaves growing over his twin jaws. Hoffa is vanishing again under the encroaching ivy. The Hydra-headed monster straddling the cesspool is doing too well; some healthy new offshoots must be cut away to preserve the shape.

During the long months of January and February, there is that perennial anguish over decisions to be made when the cold breaks. Ol' Rock, for instance. For centuries Rock's three trunks grew fatter by the year, all topped by a single crown of soaring elegance. Left alone for a decade, however, it was engulfed by ivy. Thick ropes strangled the tree from below and broad leaves blocked the sun at the top. Branches rose straight up to the open sky, like flagpoles. Meantime, shoots at the base had grown tall, sucking vitality from the top.

By the time I dug out Rock, the main tree was all but dead. Next to it, a healthy trunk, thick as a thigh, rose from the root mass; it produced huge amounts of plump olives. If I was lucky, brutal pruning would encourage the struggling old limbs to sprout new buds. In time, small bunches of green would grow into healthy new branches, restoring the tree to its former splendor. But what if I was wrong and the crown was too fatigued to rally? The old and new would both be gone, and I'd be starting from only the root.

I bit the bullet and cut away the young trunk. Reluctant to trust fate, I compromised on the top. Instead of sawing off the ugly uprights, I cut them halfway to leave some tufts of green at the tips. Then I dug up a wide circle of ground around the base and dumped on fertilizer. Sure enough, in spring the shorn limbs sprouted buds. There were so many that none managed substantial length; they clustered atop the poles in thick clumps. Over several years I could thin them out and reshape a decent crown. In the meantime, Rock was renamed Broccoli.

In the process of such labors at resuscitation, I discovered some basic truths about tending olives. You have to love them; it is hard work, requiring attention and that essential act of faith. But there are probably no living things, Border collies included, which display such gratitude for the slightest kindness.

"Olive trees respond to man, they interact," Nicolas Maileander told me one morning. Smiling slyly, speaking in low, loving tones, he might have been talking about a mistress. "There is no more passionate tree anywhere, nothing that relates to man like an olive. Just scratch around their bottoms, and they show their thanks on top. If you prune them hard, they repay you almost immediately, coming back fresher and fuller at the first good weather." He repeated proverbs I would hear again and again, words that the old Provençals put into the mouths of their trees as they conversed with them.

Maileander was my first mentor, a respected elder in the highly heterogeneous tribe of olive people, millions strong, which I would find scattered around the world. Dapper, with metal-framed glasses, he might be composing essays at a café table up north in butter country. But he has grown trees all his life. Until 1970 he made his living with twenty acres of apples. "Someone told me olives were easier," he said, "so I tried them. Smartest thing I ever did. All you do is dig around them a little bit a couple of times a year. You prune them once a year or so. That's it. They're always green, even in winter. And they are so beautiful."

Today Maileander's nursery sells five to ten thousand trees a year. Bud cuttings are planted in special soil until they sprout roots. In winter, they are potted and left to grow. An elaborate system of tubes waters them, drip by drip. At two years, eager buyers snap them up. Local farmers carry them off by the hundreds to replace plowed-up vineyards. Parisians with second homes buy them for the garden.

By the mid-1990s Maileander had edged toward retirement, leaving the nursery to his daughter, Sylvie. She clips buds, transplants young shoots, and waters seedlings as if they were treasured pets. Inquiries from abroad suggest new markets, but Sylvie is dubious. Olive trees, no mass product, need personal attention. One morning I found her in something near shell shock. A fax from Iran had asked for the price of a million trees.

The Maileanders sell picholines, the prized French variety which produce oblong olives that are best cured green. But their specialty is the bouteillan, a fast-growing tree with olives that mature to a round, plump black. Their bouteillan oil regularly wins medals at the Paris agricultural fair and ranks among the world's best. It typifies our neighborhood oil: pale yellow, fruity with a slight bite, mellow but versatile enough for anything from steamed beans to grilled fish. The bouquet only subtly suggests olives; some people taste fig. The texture is substantial but not heavy. When filtered, its surface flashes gold in the light. But most of us like it cloudy with tiny bits of the fruit.

"The olive is an amazing plant when you think of it, immortal like nothing else," Maileander said. Not far away, in Aups, groves of trees are believed to date back to roots planted when Caesar's legions built, from Paris, yet another road that led to Rome. "You can take a tree that has been completely abandoned, covered over by forest and forgotten for decades, and bring it back to full life. All you need to do is clear away around it, cut off the dead wood, and it springs back."

True enough. For living proof, I needed to look no farther than those trees entrusted to my neighbor, old man Romana.

Jean-Antoine Romana does not want anyone to notice but, nearly ninety, he is slowing down. Once I found him in the fields raking up cut hay. Bent nearly double from ossified joints, he was leaning on a cane and pulling himself forward, step by painful step. With his free hand, he dragged a rake behind him. Slowly but steadily a pitiful pile collected at the end of the row. When he rested, he stood as still as a stuffed figure meant to scare off the birds, in threadbare tweed motoring cap, faded sweater, blue peasant pants, and zip-up rubber boots.

A few months later he was out there helping his son, Jeannot, harvest the olives at impressive speed. Winter winds had blown down two-thirds of the crop, especially the red caillet roux, which drop readily when ripe. On his knees, Romana snatched up firm fruit from the ground and tossed it into a basket.

The Romanas live in Ampus, my village, which is as good a place as any to understand the troubles facing Mediterranean farmers. It is in the Haut Var, up the mountain from the little city of Draguignan, which is due north of Saint-Tropez. The winding road passes a turnoff to our steeply sloping olive groves and then bends downward sharply into the valley. The panorama is a wilder Tuscany: the same pastels, vivid greens, and old stones, but with distant mountain peaks that whiten in winter. A hilltop cluster of medieval buildings perches on the horizon.

Every direction is a Cézanne vista, of wild rosemary and lavender, brooms called genêts, and wildflowers too numerous to name. In early summer, before the big heat, when all juices are flowing, it is red on the ground from the Monet poppies and buttercup-yellow higher up from the genêts. The air is rich with the scents that perfume makers down the road capture in their tiny bottles. During those gentle months, the olive trees are sated with moisture, and their leaves flatten and turn out for better transpiration. This is when they flash silver in the sun. For old-timers like the Romanas, early summer is only the cream on the *tarte*. They love it in every color, cold and wet, or oven-hot.

The Romanas' farm is just outside town, a tidy ensemble of tractors, chicken coops, barns, haystacks, and gardens. Life revolves around a glassed-in sunporch and a small dining-room table that serves equally for the morning bowls of coffee and the holiday goose. When I first visited, I found old man Romana, his wife, Francesca (now, Françoise), and Jeannot gathered around the remains of lunch, watching the noontime news on television. Irritated growers were flinging apples at the cops trying to restrain them. The Romanas stared at the screen as if they were seeing the Grim Reaper.

This was at the height of a trade war between the United States and the European Union, early in the 1990s. France was the villain. American negotiators railed against European subsidies, overlooking the fact that Georgia peanut growers were probably the most artificially propped-up farmers in the world. The French, and others, argued that more than prices were at stake. If small farmers could not survive against competition from big-time agribusiness, entire rural areas would depopulate and a thousand-year-old way of life would vanish. The Romanas were Exhibit A.

Romana grew up on a farm in Piemonte. When Mussolini's oppression was added to a hopeless economic situation, he decided to take whatever he could find across the nearest border. He reached Ampus in 1938, with a wife, a cardboard suitcase, and an accordion. At first the young couple tended sheep, running their flock on the high flank of the mountain above Draguignan. Then they worked out a feudal arrangement, which suited them well for fifty years.

The mayor of Grasse owned a large grove of abandoned, overgrown olive trees up where the sheep grazed. If they could make them produce, half the crop would be theirs. Several inheritances later, the deal still stands. The Romanas later bought a farm as well. But whatever else they produced, those olives were their life—and the family's mainstay.

Now Jeannot, who has worked from dawn to dark since he was a little kid, just shakes his head when asked about the future. His parents have already sold the farm they had built up over decades; they've retrenched onto a smaller piece of leased land. He and his mother go each Saturday to sell vegetables at the Draguignan market. The price of leeks is constant. Tractor prices skyrocket, year after year, along with fertilizers, pesticides, and everything else that does not grow out back.

"I don't know if we can make it," Jeannot told me. "We've already sold off the cows. Wheat, potatoes bring in a little money. Not enough. I'd like to get married, but in no way can this farm support a wife. Besides," he added, with a rueful chuckle, "when am I going to have time to find someone?"

In these straits, growing olives is a luxury, a simple act of love. If the year is very good, the family might produce a thousand liters of oil. In the 1995–96 season, however, the total was three hundred. Adding the European Union subsidy but taking out expenses, that comes to something near three thousand dollars—before the trees' owner gets his share. In peak years, the money goes up, but that means picking five metric tons of olives, weeks of all-day labor.

The family does all the work. Sometimes Françoise's niece from her old home in Piemonte comes to help. Other relatives pitch in, if they have the time. Hired hands, however, are out of the question. "If we paid someone, he could not pick enough olives to cover his wages," Jeannot said. "It would be cheaper to buy oil."

Still, in every season, unbelievably good smells arise from the omni-purpose table, and the sunroom exudes high spirits. Françoise, well into her eighties, is usually stirring or sewing and always cracking jokes. Friends and relatives stream in for hot coffee or counsel. Jeannot, with a handsome round face, electric hazel eyes, and a peaceful mien, laughs a lot for a suffering son of the soil.

Old man Romana, hard-of-hearing, sits impassively in his favorite chair and puffs a hand-rolled cigarette. But anytime a visitor manages to say "olive" loudly enough, he springs to life. A grin exposes the gaps in his teeth. Eyes gleam behind glasses as thick as the bottoms of wine bottles. With a young man's vitality, in Italian-tinged French, he spouts a stream-of-consciousness recitation of memories, statistics, and advice.

"Very important, never cut the ends," he told me once, repeating this several times for emphasis. When pruning or picking, he meant, do not damage the branch tips from where the next year's growth must come. "You've got to take care of trees, fertilize." At that, he shifted focus, con-juring up an image of his own trees growing near mine, high up the distant hill. "They're beautiful, aren't they? The ones by the road aren't so good. But farther down . . . Those are trees."

When he first came upon his trees, the hillside looked like Cambo-dian rain forest. Oaks had taken root next to the olives, eventually dwarfing them. Thick, bushy genêts filled in the empty spaces. Over time, leafy vines wound up the oak trunks and on into space, blocking off the air and light that olives badly need. Under this lush canopy, wild berry brambles wove themselves into fierce and impenetrable hedges. Everywhere, ivy sucked moisture and nutrients out of the soil. Ancient stone terrace walls, weak-ened by wayward roots, collapsed with neglect. Here and there a telltale green-gray olive branch poked out above the tangle in a desperate search for sunlight.

"My wife and I, we worked like animals to clear the land, from the first light of day until it was too dark to see," Romana said. "I cut, and she hauled out trees with the help of a donkey. We dug out oak trees, dead stumps, everything that did not belong. And then we rebuilt the *berges* [terrace walls] so the trees would catch the rain that fell."

The Romanas were learning on the job. They had farmed in Italy but knew nothing about olives. With some advice from neighbors, they cut back the trees, sawing away the long, woody limbs that had risen like periscopes to find light above the engulfing vegetation. They dug up the earth in wide circles around each tree to weed out encroaching plants and aerate the soil. Day after day, they scooped up the droppings of their sheep to feed the olive trees.

This was enchanted country. Ancient Greeks explored it, but the Romans settled it, building dramatic arched stone bridges over ravines on their way to the Rhône and the Seine beyond. Everywhere, they planted olives. A thousand years later, local nobles erected a fortified castle, which is now a ruin on the steep slope across from Romana's grove.

Marie was born during the clearing operations. Lucie and Annie followed. By the time Jeannot was born, in 1948, the trees were bearing fruit. With brute force and blind faith, the Romanas had brought historical monuments back to life. Old hands now, they were grafting branches to improve yield and quality. Their resurrected trees were among the best in groves which stretched for miles above and below the winding mountain road. The family had made it as immigrants, happily installed in an old stone house in the midst of their flourishing olives.

In those early years the work was constant. All healthy trees throw out shoots from the root ball; these must be stripped away. After severe spring pruning, a waist-high skirt can envelop a healthy trunk by fall. But when an abandoned tree is brought back from near-death and fed with fertilizer, it throbs with extra energy. Shoots grow so fast you can almost hear them.

The crop was good in 1955, and Jeannot was old enough to help with the harvest. After harvesting and pressing, they sold their oil and hunkered down until the cold passed, so they could prune for yet another good year. Warm temperatures in January deceived the olives. Sensing an early spring, the old trees drew sap up from the roots and began working on buds. And on February 1, in a matter of hours, the temperature plummeted below −7 degrees Centigrade. Trunks froze and exploded.

In that winter of '56, there were tears across southern France. Tough old *paysans* surveyed their cracked and lifeless trees, and they wept. Six million French trees suffered, a third of the total, and a million died. For olive people, it was the worst calamity since record-keeping began in 1739. In the spring, whole groves were uprooted—surviving trees included—to make room for grapes. A few isolated pockets escaped the worst of it, but the Romanas' orchard was not one of them.

"We lost heart, just gave up," Romana remembers. He told the owner of the grove that he was finished with olives, maybe even with France. The owner, disheartened himself, sold his main piece of land and the stone house, but he kept a separate patch of 150 trees. Romana moved to a farm in the nearby valley. Maybe he wasn't done with the olives yet.

He sawed the lifeless trunks down to the ground. Over the next few seasons he tended and staked the strongest new shoots. Now, forty years after the cataclysm, his trees are, as he says, beautiful. Many are still ancient sculptures, their leathery green leaves catching color from the sun. Others are made up of four thinner trunks around massive stumps, each with five hundred age rings whorled tightly in a natural work of art.

Along our little road, the old-time growers are as gnarled and picturesque as their trees. Over generations, the curious French property code has produced numerous permutations. Not only the land but also the buildings are divided into tiny, odd-shaped parcels. When I first looked around, an eager seller offered a fine old house and six acres. Except, of course, for the room in which *le père* Colle kept his tools.

Colle, somewhere between irascible and cantankerous at eighty, owned much of the hillside, including that room. He roams the ridge at his pleasure. Once, wind toppled an old oak tree onto the property of a Belgian newcomer. The man called for help in sectioning the tree, which bisected his front lawn. Within an hour, Colle was there in full fury, demanding his rights. Under the law, no one could touch the tree until property owners in the immediate vicinity met to determine how to divide up the timber.

Old Ricard is about seventy and milder-mannered. He simply wants to be left alone to tend his trees. One afternoon I watched him jamming a crowbar under a stump that was much bigger than he was. He had spent most of a week digging out the mammoth root, swinging a pick for hours on end. A month later, I saw a modest young tree in its place. The thought that he might not be around to harvest the olives it would eventually produce did not seem to bother him.

The cast of characters changes with the years. Everyone knew Grand Daniel, Big Dan, who lived far down the ravine without even a tractor path to link him to the wider world. Each January he carried his olives up the hill on his back, a hundred pounds at a time. A loner, he died in his cabin, and it was weeks before anyone thought to go look for him.

Now neighbors talk about the bourgeois Parisian, a part-time resident who fires up his chain saw to cut a half-inch rose stalk. Or the famed French cartoonist and the quirky American journalist who, when visiting the former, fell for the place.

But Romana, with no title to anything up on the ridge, is king of the mountain. If there is anyone to rival his position, it is Roger Martin, whose own beautiful trees are just across the narrow road, uphill. Getting on in years, he scampers up his trees like a squirrel, pruning branches with a sure hand at lightning speed. His paying job, until he retired, was looking after four thousand olive trees at La Treille, the largest grove in the area. Martin is Romana's son-in-law, married to Lucie.

The dynasty has deep roots. Martin's son, Eric, has his own fine trees, and two kids growing up among them. He took over from his father as foreman at La Treille.

Just past my trees, another impressive grove belongs to Paul Bosquet, who works down in Draguignan but spends the weekends pursuing his passion. His wife is Marie Romana. At the edge of Bosquet's grove is a crumbling ruin, high stone walls with no roof or door. It is the old Romana home, abandoned in 1956. Bosquet and his wife have revived the trees. Olive trunks might die when the sap freezes, but the roots usually survive. The *Olea europaea* is one tough species. On Bosquet's trees centuries-old grafts survived the cold. Where others didn't, he grafted new species. One giant tree is part caillet noir and part caillet roux. On one half, the ripe olives are black; on the other, they are red.

One Christmas I noticed that the Romanas' trees looked a little un-kempt. They were skirted with tall shoots that someone should have ripped away months earlier. Olives were still on the branches, and the mills would soon close down. I stopped by the farmhouse in the valley to see if anything was wrong.

Jeannot greeted me cheerfully. No problem. Everyone was just busy. His mother was frantically stitching up an enormous sheet of plastic mesh to help anchor the sheep shelter roof. Mistral winds had carried off two earlier roofs. Another twenty or thirty other things had to be done, but no one would forget the olives.

The last time I'd visited, Françoise loaded me up with enough vegetables to make soup for twenty-five starving teenagers. This time I delivered a jar of my chili. It is hot, I warned, a little anxious about octogenarian digestive tracts.

"How hot?" Françoise asked.

"Very, very hot."

"Good."

Françoise has not forgotten a single detail of her life since World War I broke out, and she knows a lot about olives. She is also a reasonable sit-down comedienne.

I mentioned that the oil was a little earthy this year.

"Worms," she said.

I grimaced.

"Oh no, worms are good. They add meat to the oil."

She was more serious on the subject of Jeannot. His bad back needed medical attention and a good rest. But that did not fit in with his twelve-hour day of lifting rocks, manhandling a plow, and chopping wood.

"He's used to working like that, started very young," she said, shaking her head. "I wish he could stop, but how can he?" It was late in the day, and Jeannot's buddies were already knocking back pastis at Les Braconniers in town. They'd be loaded and laughing by the time he sat down to dinner and then collapsed into bed. For reasons I've never been able to fathom, however, Jeannot is always smiling, and so is his father.

I met another member of the family. Bosquet's son, Bernard, had just come back from the army. Restless, curious, he was ready to see more of the world. He peppered me with questions, the only person within miles who ever seemed particularly interested in events down the hill. Maybe he'd study law. Maybe he'd pilot an airliner. But, I asked him, would he continue to tend the family olives? Bernard looked at me as if I were a little weird. "Of course," he replied.

Up near my place I often walk to Bosquet's orchard to marvel at what I call the Romana family tree. The ancient root ball continues upward to form a massive triangular trunk. It is shaped like Plymouth Rock and is nearly as big. Also, it is probably just as hardy. Perhaps nuclear war might kill it, but the 1956 frost did not.

Wild Olives came with its own resident character. He is not thrilled with publicity, so it might be best to refer to him by the local nickname he seems to treasure: *l'emmerdeur du coin*. The Neighborhood Pain-in-the-Ass. This sobriquet comes from his zealous vigilance: every tradesman, hunter, old friend, Boy Scout on a hike, vandal, or meter reader is likely to encounter his hulking presence. In extremis, he brings his shotgun.

N.P.A. retired as a truck driver to settle in the country. A skilled mechanic, a passable planter, and a huffing-puffing Goliath with pick and shovel, he is a very good man to know. Under an exterior that can only be called grumpy, he has, yes, a heart of gold. He is there when the train arrives at 6 a.m. He waters the plants over scorching summers when I am off in West Malaria. He hauls me out of the ditch when I miscalculate the evil turnaround point on my access road. He restoreth my soul at holiday time. And he drives me nuts.

Though we are of similar age and at least equal intelligence, N.P.A.'s principal amusement is chortling about the idiot American writer who buys a different sort of tool, leaves too much vegetation uncut, eats funny food, and insists on engaging specialists—not him—to help work the olives.

My first tip-off was when he and his wife, a wonderful woman, came for lunch. They regaled me with stories about our absurd Parisian neighbor and others. I should have noticed they were glancing around my place for more fodder for their stories. It is churlish to take offense; both have been generous and welcoming, and Wild Olives could not survive without them.

When I began to experiment with varieties and oils, I mentioned that some rather gifted olive growers were sending me a couple of picual trees from Spain. N.P.A. rolled his eyes and snorted. Spanish olives? I offered him a sampling of oils from Tuscany, highly spiced and fruity. He grimaced. "No taste," he pronounced. He all but spat out a nicely mellowed Lucca oil. "It's the same crap," he said. But then the generator broke down on my decrepit Citroën, and of course, he fixed it. One must take the larger view.

In the end, exposure to N.P.A. taught me the single most important fact about olives: they are probably the most forgiving tree on the face of the earth. His trees, like the Romanas' and mine, were brought back from the dead. At every step of the way he applies homespun wisdom picked up from the quirkiest of sources.

Routinely, our trees fall victim to *fumagine*, more properly *Capnodium elaeophilum*, a fungus caused by the sticky trail of black scale insects. The bark turns a velvety black and then, left untreated, blisters away. Everyone else zaps it with a copper compound. N.P.A. uses Clorox, and it works. He explains each unusual approach with a folksy homily. Most growers turn over the soil around their trees after harvesting in December to kill grass roots before spring. He desists, refusing "to bury the cold." And his olives thrive.

Only once did our differing styles veer dangerously close to homicide. That was when he parted my jungle with the N.P.A. Freeway. One fall, too busy with real life, I asked N.P.A. if he would thin out the summer's encroachment on a leafy glade that shielded my house. Left on their own, our fields are soon tangled with brambles and vines. Ivy roots snake up tree trunks, choking them in python fashion. Genêts die and curl to ugly brown, giant torches awaiting any sort of spark to flash into flame. Oaks drop acorns, which take root in awkward places. Pines grow fast, shedding needles which acidify the soil. This extraneous vegetation is deadly for the olives, but it is also beautiful.

I try to keep a balance, protecting my trees while conserving the jungle. N.P.A. likes to slash and burn. I had in mind a few simple cuts, which would provide him with genet logs for his fireplace. When I returned, it looked as if a battalion of army engineers had taken bulldozers and napalm to the Ho Chi Minh Trail. I was furious. The olives, however, were a whole lot happier.

Between the wars, a stream of Italians left Piemonte for the hills of Provence. About the time the Romanas were figuring out how to grow olives,

Giovanni Rovera and Diodato Doleatto were learning how to press them. The two men, strangers at the time, bought old mills less than a mile apart in the village of Flayosquet, downhill from Ampus near Draguignan. Today, the families are still at it, but they are no longer strangers. Every season they glower at each other from a distance, and each will gladly tell visitors why his particular technique is better than the other's.

There are a lot of ways to press an olive. In most olive-growing countries, millers favor the modern continuous systems, which use linked centrifuges to grind up pulp and, at the end, spin out oil. The advantage is economic; these are fast and virtually run by themselves. But France makes only about two thousand tons of oil a year, three hundred times less than Spain. Like most French millers, the Roveras and Doleattos use the traditional method, squeezing olive mash in a tower press. Their battle is over the final step.

In the traditional way, olives are crushed to an oily paste by a huge granite wheel, or sometimes two wheels. The mash is then spread evenly onto round, double-layered mats (*scourtins* in French) with holes in the center. The mats are woven from natural fibers, and each is the size of a manhole cover. Millers stack these mats onto a platform, placing them over a central steel pillar. Hydraulic power then squeezes this towering sandwich, and oily liquid runs down the side into decanting tubs. Sometimes hot water is hosed onto the stack to hasten extraction by breaking up cells in the pulp that contain tiny oil droplets.

When this process is completed, olive oil must be separated from natural vegetable water in the fruit and any added water. Alfred Rovera, Giovanni's son, uses a vertical centrifuge. The Doleattos do it the old way, *à la feuille*.

A *feuille*—the word means leaf—is a slightly cupped, round iron spatula with a short handle. After the pressed liquid settles, millers slip the *feuille* under the layer of pure oil that floats atop the water and lift it into a steel or plastic barrel. Some scoops are shaped more like low-walled skillets, for greater volume. But each is wielded with elaborate gestures by a practiced hand. Swirling motions leave behind lingering drops of water. Pouring at arm's length reflects golden sparkles of light.

To compare the two approaches, I made back-to-back visits. For no particular reason, I gave Rovera the first shot.

"Ours is much better oil," Rovera declared. "Cleaner. Clearer. And it is much faster for the customer. I can press five thousand kilos of olives a day. What can he do? Maybe eight hundred? Hah." There was, perhaps, a certain folkloric advantage to following the old way, he allowed, but at what price?

At retirement age, he is lanky and gaunt. He yawned occasionally, having sat up all night with a sick boiler, but he clearly liked his work. With great affection he showed off his machine. Oil from the press decants briefly in two tiled vats, and some excess water is pushed down the drain by the weight of pure oil which collects on the top. Then the liquid—a mixture of oil, water, and olive bits—is pumped to a holding tank above the spinner. Heated water is piped into the machine, and then the oily water follows. Pure oil comes out one pipe. Waste water exits from another.

"Better," he said. "Much better. If you're going to the market, do you want to walk or take a bike? It is obvious."

He did win the blind-tasting contest at the Draguignan Olive Fair in 1994. But "better" and, certainly, "best" are relative concepts among olive people. A single batch of oil can be entered for a prize, and the rest may not be as good. At the very least, he knows his profession.

Rovera's father bought the mill the same year that Romana brought his wife and his accordion to Ampus, up the hill. He had been coming to Flayosquet to work each winter since 1928, returning home to Piemonte for summers to help his father in the fields. In 1938 he scraped together all his savings and bought the mill. That was the year the war broke out. Alfred, then eight, stayed in Italy with his mother, while his father was in Flayosquet. France refused to let them across the border.

"He ran the mill with the help of some other men, and he did fairly well," Rovera said. The war, he added, offered opportunities for astute businessmen who supplied such a valuable commodity as olive oil. In 1946 the family was reunited. The mill flourished, and the 1955 crop was good. Then in 1956 there was the freeze.

Until the cataclysm, dozens of oil presses dotted the Var, many of them turned by water-driven wooden gears. Most of them closed during the years when there were few olives to press. The Roveras nearly went under with the rest, but they persevered, and the region recovered. The old man died in 1978. Alfred renovated the mill and installed his machine.

"It is the only way," he said, patting the gleaming stainless-steel cylinder. "I would guess that 99.99 percent of the olive oil around here is separated by centrifugal force."

Not exactly. Besides the Doleattos, a hearty band of purists still like to do it the old way. Oil is separated *à la feuille* at both mills in Draguignan; Fabrice Godet offers a choice. Eugène Mauro, who was running his mill when Godet was born, still only does it by hand. Our ancient mill at Tourtour uses *la feuille*.

In fact, not even the Rovera family is unanimous. I mentioned the *à la feuille* business to Madame Rovera, who helped at the mill. She had not heard my conversation with her husband. I told her that Godet was disconsolate that the majority of his customers preferred that he run their oil through the centrifuge. He felt the commotion of the machine emulsified the final product. But people liked to see through their oil.

"That's how it is these days," she said, nodding agreement with some vehemence. "People want everything quick, easy, too clean. There is no appreciation for real quality anymore. Look at the supermarkets. Someone will pay twice as much for a bright red apple that will make them sick instead of buying something that looks like a real apple and tastes better."

The different styles did not seem to affect anyone's business. Except in bad years, Rovera said, there was plenty of work for everyone. A severe freeze in 1985 did serious damage, but most trees came back. And, since about 1990, more olives are being planted every year. As the demand for oil grows, farmers and homeowners are ripping out grapevines to plant trees. The regional government offers a subsidy to encourage them.

"People are realizing the fact that in the Var grapes grow better by the coast and olives are better higher up, farther from the olive fly and all that humidity," Rovera said. He offered a last bit of wisdom: "Just remember. Always buy your wine south of Highway 7 and your oil north of it."

On the way out, I broached the Dark Rumor. A neighbor of mine had muttered that he stopped going to Rovera because he suspected that the miller cut his product with peanut oil. This did not seem very likely. Another neighbor, whose oil is excellent, swears by Rovera.

A local expert had a different story. It was not peanut oil; that mixes badly with olive oil. Instead, he said, Rovera cut his product with cheaper olive oil bought in bulk from North Africa or Spain. Obviously the rumors had gone around for some time, because Rovera was ready with an emphatic denial.

"See all of this?" he said, gesturing to rows of containers by the door. "It is inspected, regularly, carefully, by the authorities. It would be impossible . . ." I stopped him. In fact, oil is not inspected all that carefully, or regularly, by the authorities. Even if a miller is cited, fines are too low to discourage any profitable shenanigans. I had trouble imagining anyone going through all that labor of love to pollute the final result. But who could know.

In France, suspicion of oil millers is hardly new. In *Jean de Florette*, Pagnol offered this passage about his hunchbacked tax-collector-turned-peasant: "The old trees, cleared of offshoots and dead branches, and pruned haphazardly to feed the goats, had been grateful for long-forgotten attentions, and they gave thirty-five liters of oil, since the honest miller at Bramafan, moved by Monsieur Jean's kindness and hump, skimmed off only 10 percent."

As we said goodbye, as long as we were into the blunter questions, I asked Rovera how he got along with his nearby rival, Doleatto. "Oh, we say, 'Bonjour,' when we meet, but I would not say we're friends," he replied. I pressed for details. "Well, let's say I do not believe in criticizing my confrères, but apparently he is not of that opinion."

Up the road, Noël Doleatto did not beat around any bushes. "No, we're not friends," he said flatly. "I believe you should say things as they are." He did not elaborate.

Noël's great-grandfather bought the mill in 1924. Noël's son, Max, took it over in 1988, while still in his twenties, and institutionalized the old way. As Max sees it, when a mill has operated successfully for four centuries, why destroy tradition for something worse? He replaced the old wooden screw with hydraulic power. But that's about it.

Max quickly established his reputation. Each year he is on the jury which picks France's best oil at the Paris Agricultural Fair. People order his oil from all over the country. A good-natured sort, he just laughs at the mention of Rovera. Plainly he does not regard him as a credit to the profession.

"Of course he says that his way is better," he said. "He has to come up with reasons to justify an easier and more profitable way to produce oil." But Rovera was wrong, Max maintained; he could press fifteen hundred kilos a day. That was still a third less than Rovera, and it was damned hard work. But the customer got much better oil as a result.

"The hot water alters quality," he explained. "In the centrifuge, molecules explode. It is too violent. If you let the oil settle for long enough, it comes back to normal, but the fruit isn't there anymore. The flavor is off."

I was not going to get into the feud. As for the oil, however, it seemed open and shut. No one was going to spin my olives. Rovera's product was good, but the Doleattos' was much richer in those delicate, indescribable nuances that make an oil great. The taste and aroma of oil depend on highly volatile flavor compounds released in pressing. Heated water added to a centrifuge can wash them away. Friction from high-speed spinning can damage them.

Of course, I was prejudiced. I liked the taste and character of the cloudy, natural stuff. Faster production was an irrelevant matter of oil-mill economics. Three thousand years of tradition had to count for something. True, hand-separated oil must be decanted carefully or impurities left in it might go rancid. The trick is finding someone who does it right.

Max runs a very impressive operation. Like other millers, he requires growers to deliver a minimum of 240 kilos of olives if they want their own oil in a separate pressing. Otherwise their olives are mixed with other odd lots, and they will get a proportionate number of liters. At the same time, Max presses his own olives and sells the oil in bulk from large containers.

You can get a sense of a mill by the way finished oil is handled. The Tourtour press is by far the most picturesque. The crushing wheel is powered by water, driven by rough wood-peg gears, and the press is cranked down on a carved wooden screw. Its oaken beams are enormous, set in medieval stone. Oil, however, is left to settle in green plastic garbage pails. "We keep it for two or three days," remarked Loulou Dauphin, the mill hand, who looks after village streets when it is not pressing time. "It'd be better if we decanted for five days, but that would mean fourteen barrels, and it's enough of a *bordel* here with seven."

The Doleattos decant their customers' oil for two weeks in tidy pale-green metal casks, properly lined, with tight wooden lids. Their own oil sits for a month. When ready, they dip it out with zinc-plated beakers until it is almost all removed. The last few inches are lifted off with the *feuille*.

Among the casks sits a single waist-high clay amphora, glazed on the inside, from a much earlier time. "We used to have a lot of these, but this is the last one left," Max said. "My grandfather didn't shut down after 1956, but he survived only by selling off the jars."

Max produced some plastic teaspoons, explaining that metal alters the flavor. A professional taster, he laughed off the formal Italian method. He lets oil reach the taste buds at the back of the tongue and waits a moment before swallowing. Even when judging thirty samples in Paris, he spits out the oil only if it is disgusting. He uses bread to clear his palate. He does not suck, smack, or slurp. "All that noise some people make, *ça c'est plutôt du cinéma*," he said. *"Ça sert à rien."* It is showing off; it has no purpose.

From separate barrels, we tried oils made from caillet roux, the willowy *plante de Figanière*. The first batch was from olives picked at the beginning of December. It was, in a word, incredible: round, fruity, with that pleasant Var sweetness but also a formidable Tuscan kick. The second batch of olives was collected at the end of December, allowing a month more of ripening on the tree. It was also delicious: mellower, fruitier, with a buttery feel and less of an afterbite.

I suggested a blend. Max mixed up some half and half, dribbling out the last drops with the concentration of a perfume maker and an amused patter: "Let's see now, two bat wings, a newt's tongue . . ." I tasted. *"Fais le plein,"* I told him, producing a five-liter demijohn and three more assorted bottles. Fill 'er up.

Max took his time, in the way of an old-time tradesman, but he was a little harried. Three people were waiting with loads of olives. Two more had come for their oil. A few tourists were poking around, and someone from the village wanted a tour for visiting friends. Mainly, however, the mistral had set his chimney on fire, and the place was full of smoke.

"Well, there are minor drawbacks to tradition," he said, hurriedly stuffing things into the various cracks and openings of his flue to stop the draft. The strong wind had created a blast-furnace effect, igniting the carbon cake inside the chimney of the boiler he used to heat water for spraying the *scourtins*. It would shut him down for hours, if not longer.

"The old guys set the piping deep into stone and cement," Max said. "It's a real mess to get to when something goes wrong." Compared to what else usually went wrong during an average season, this was a minor calamity. Even when everything goes smoothly, it is killing labor. The eighty *scourtins* take two days to clean thoroughly after repeated pressings. Residue left in the fibers can turn rancid and ruin fresh oil. The mill must be kept scrupulously clean. The water-powered wheel is temperamental; gears and axles break.

"It's in my blood," Max said. "I was born on a pile of olives." Not exactly, but close. As a kid, his parents left him to play atop an enormous mound of ripe olives awaiting the press. He started eating them. The bitter fruit made him so sick that it was years before he could face another olive, no matter how delectably prepared.

I asked if the difficult, repetitive work did not sometimes get him down. "No," he said, with a gesture that needed no translation: That's the olive biz.

If a miller is in a good mood and your timing is right, he'll hand you a *bougnotte*, which rivals truffled foie gras as a French delicacy. Toasted bread rubbed with garlic, it is sprinkled with salt and splashed with fresh olive oil. Around Nice, it's a *brissa*. Italians will recognize it as *bruschetta*. All that is different is the flavor of the oil.

In French olive country, nearly every cluster of villages has its special varieties and at least one mill nearby to press their olives. The list of varieties is long, partly because the same olive often goes by several names. Among the most prized for oil are bouteillan and aglandeau. The caillet roux has a wonderful taste, but it grows more slowly and suffers in the cold.

Nyons, far to the north, has its tanche, round, fat, brown, and tasty, rich in mild oil. The salonenque is found north of Marseille, especially around Salon. It has a high content of the buttery but slightly spicy oil pressed in the valley of Les Baux. The grossane shares the same region; it is a large, firm table olive, cured black in salt or brine. Farther west is the lucques, France's most prized green table olive.

Around Nice, cailletiers make not only fine oil but also those little black olives sprinkled in a *salade niçoise*. On the Côte d'Azur and our Var highlands are different sorts of caillets (or cayets), each with a distinctive flavor. The widespread picholines, grown all over French olive country, produce excellent green olives for eating, but their delicate oil is fruity and sharp. And there are others.

"Ask anyone and he'll swear that his oil is the best," Jean-Charles Aycard said with a hearty laugh. "The French are like that. It's all taste, preference, and a little luck." He thinks the hand-separated oil he makes in Salernes from local subvarieties is better than anything else he has tasted, but he doesn't boast about it.

Aycard, a retired navy man, runs the cooperative press mostly on a volunteer basis, earning little. "I do it because I love it." He presses only olives grown in the immediate vicinity, with an occasional load from Villecroze, five miles away. And he is ruthless about what olives he accepts.

"That is the only way to get consistent quality," he insists. "If you mix in olives from different places, especially if you don't know the growers and the trees, you're never sure what you will get. And if any of the olives are wormy or rotten, you might as well toss out the oil."

Aycard snorted in kindly derision when I explained my goal of producing enough olives for my own pressing—the famous 240-kilo minimum. "You'll get the oil from the guy before you," he said. "The *scourtins* are soaked so thoroughly that they create a sort of pipeline. What you put in the top is not what you'll get at the bottom. To be sure, you need two loads, and then you can only count on the last five or ten liters being yours. You can try to be the first of the season, but you'll only get a few liters, and it'll taste awful. Try mopping water with a dry sponge. It's the same principle."

He had more bad news. With the European Union's mania for standardizing, and its penchant for big business, he believed the artisanal oil press was on its way out. Already, he saw the handwriting all over his old stone walls. The Brussels Eurocrats would not have to outlaw any process. A few more regulations about chemical component levels or somesuch, perhaps combined with subsidy changes, and *à la feuille* would go the way of artisanal English sausages and German beer.

This would be a tragedy. Of the world's endangered species, old-time olive millers have to be among the most fascinating. There are, for instance, the Godets in Draguignan.

Lucien Godet, white-haired, with the impish air of an underfed Santa

Claus, leaves the garlic off his *bougnottes*. He doubtless tasted his first well before starting kindergarten, and oily toast has been his staple every day since. From the way he smacked his lips and watched me wolf down the crust he gave me, you'd have thought he had just invented a brand-new treat.

Godet built the water wheel that drove the abandoned mill in Ampus. He has been growing olives, or milling them, for nearly sixty years. His son, Fabrice, runs their Moulin de l'Horloge, a Victor Hugoesque stone barn with gold-medal certificates tacked to the massive wood beams. Fabrice has the curly black hair and sloe eyes of a rock musician. He wasn't sure what to do until he tried the olive business eleven years earlier. Now he is in it for life.

Fabrice and his father talked olives well into their lunch break, a sure sign of zealotry in France. They were worried about disappearing traditions. A centrifuge separator hummed next to the tile *à la feuille* basins. "It's depressing," Fabrice said, "but most people want to spin these days. They think clear oil is better. The old-style peasants are vanishing. Today it's a different clientele."

The busiest mill within reach of Wild Olives is the Gervasonis' operation at Aups. Growers drive in from as far away as Marseille, two hours each way. Most of the Romanas' crop is pressed there, along with Maileander's and the olives that Martin brings from La Treille. Papers at the town hall mention the mill in 1750, but it was probably centuries old by then.

Charles Gervasoni, in his late eighties, still sounds as if he just got off the train from Italy, but he was born in France, near Carcassonne. He ran the mill in the old way, crushing olives with a water-driven wheel and wooden gears. When his son Jacques took over, he made the old apparatus into a corner museum. He still presses with *scourtins* in a tower, but he has replaced the old coconut-fiber mats with ones made of polypropylene.

Jacques looks less like a miller than a Cinecittà version of a polished prime minister, graying and distinguished at fifty. But, he said, he never wanted to do anything other than grow, press, and eat olives. He lamented the change from *feuille* to centrifuge but had no choice. "We've got too much to do around here," he explained. The mill turns out thirty thousand liters in a good year, not much for southern Spain, but a hell of a lot for the Var. He has taken mechanization a step beyond the others: a rudimentary Flintstones' robot lifts the laden *scourtins* onto the press.

The Gervasonis' mill presses thirty, maybe forty different varieties of olives. Until the calamitous freeze of 1929, only a few stalwart types of olives grew in the region. "We used to see the ribier, an enormous tree that could produce 150 kilos a season," Jacques said. "They were hammered in 1929. In 1956, most of the last ones went."

After the mill, Jacques's passion is a small grove he bought near the old site of Aups which the Romans settled. It was overgrown and had to be saved. He cut down a thousand-year-old tree that was so fat it took three people holding hands to encircle it. Old age and too many hard freezes had taken their toll. When he told the story, Jacques shrugged. "It was tragic," he said, "but that's nature."

The Gervasonis often replied to questions with a shrug. With olives, who knows. I had noticed this everywhere. One old-timer offered advice with iron certainty. Another, with no less self-assurance, said the opposite. And this was only France. There was a lot to learn, and it was time to get on the road.

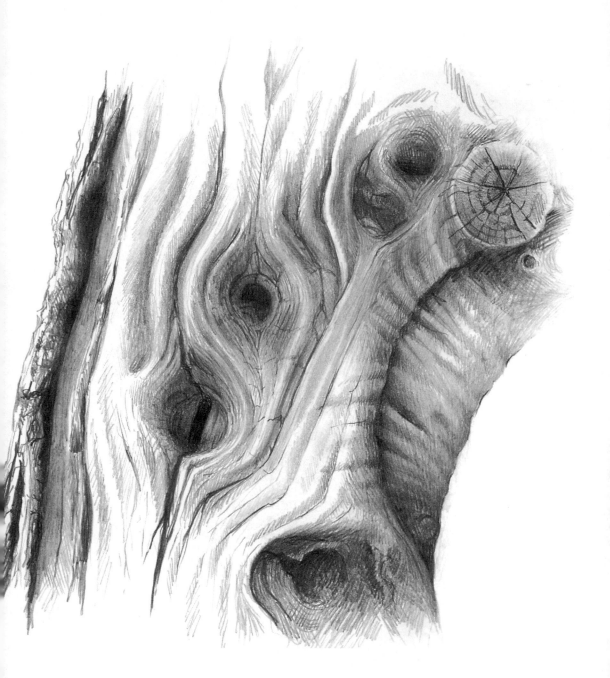

Olive Log

Jaz Mussakhkhan

To be served this elaborate chicken dish with sumac is a high honor for Palestinian guests. Habeeb Salloum and James Peters offer this version in *From the Land of Figs and Olives*.

1 3–4 pound chicken, cleaned and cut into pieces
6 cardamom seeds, crushed
¾ cup olive oil
4 large onions, chopped
Salt and pepper to taste
½ teaspoon allspice
¼ cup pine nuts
½ cup sumac
4 small loaves flat Arabic bread

Place the chicken and half the cardamom seeds in a pot and cover with water. Bring to a boil over medium heat, and cook about 45 minutes, until the chicken is tender. Remove the chicken pieces and set aside. In the meantime, in an uncovered saucepan with ½ cup oil, cook the onions, salt, pepper, allspice, and remaining cardamom over low heat for 1½ hours. Separately, sauté the pine nuts in the remaining oil until they begin to brown; then add the nuts and sumac to the onions. Stir and allow to cool. Split open the bread and place the halves in a greased casserole; divide the onion-sumac mix in half and spread one quarter of it on each layer. Top evenly with the chicken pieces and cover them with the remaining mix. Cover with foil and bake 40 minutes at 350 degrees in a preheated oven. Serve each chicken piece on a portion of bread.

Serves 4–6.

*"Allah's light is . . . as a lamp kindled
from a blessed tree, an olive, of neither the
East nor the West, whose oil is of such luminous
glow that it seems to shine though no fire has
been touched to it. A light upon light."*
—Koran, *Surah 24*

*"Happy are those who fear the Lord and walk
in His ways. Thy wife: a fruitful vine inside the
house. Thy sons: olive plants around the table."*
—Bible, *Psalm 128*

3

Mount of Olives

Not much is older than the view from the hillock overlooking the ancient seat of the Holy Land. From the Mount of Olives you survey the ageless walls of Jerusalem and the gates crashed by every manner of Hebrew, Muhammadan, Crusader, Turk, and tourist. The Dome of the Rock stands defiant—the third most sacred site of Islam—next to Judeo-Christian shrines. In some ways, nothing has changed for a thousand years. But a modern pilgrim now can use a mobile modem to fax a prayer to the Wailing Wall down below. And the olives are mostly gone.

True enough, ancient orchards and new clumps of seedlings bloom from the Mediterranean coast to the Dead Sea. Around the Sea of Galilee, they are of thrilling beauty. At the edge of deserts, their crowns offer shade and refreshing green. Late each fall eternal trees hang heavy with fruit. By the New Year the old oil jars are full again.

But the rich old groves that climbed heavenward from the Garden of

Gethsemane have been cut to a few trees, in clumps or alone, among new housing projects, a Mormon university, and the sort of suburban fungus that accompanies a society powered by a more combustible kind of oil. If jarring, this is only natural. Where its roots grow deepest, the olive is an imperfect symbol. No harbinger of peace, the Holy Land olive symbolizes instead hard struggle and enduring conflict.

Having decided to track olives to their source, I might have gone first to the Temple of Knossos on Crete, where Minoan kings shaped a civilization around their beloved oil. Or the Anatolian plain, where Turkey fades into Syria. But Jerusalem seemed a more logical place. Here, history never slipped into the past tense.

Down twisted stone lanes inside the walls, grizzled Arabs sell some of the world's best oil, green-gold nectar from the West Bank, pressed from odd-shaped little olives grown on great bushy trees planted by forgotten ancestors. Off King David Street, restaurants offer cured black souris, manzanillos, and fatter kibbutz varieties, triumphs of Israeli agriculture.

And no trees are better known than the gnarly ents in the Gethsemane churchyard, the backdrop for Jesus' bitter last night. In fact, *gethsemane* means oil press, and that famous meal may have been served under the vaulted stone of an olive mill at the grove.

On a scorching summer morning I appeared at the gate amid a busload of Orange County sightseers. Soon afterward, a man named Mahmoud caught me peering with rapture at a leafy double-trunked statue of boles and burls. He had a boxy camera around his neck and a pocketful of cards announcing his profession: guide.

"How old are these trees?" I asked.

"Five thousand years, maybe more," he said. It must have been my Jerry Jeff Walker baseball cap.

"Seriously please, Mahmoud. How old are these trees?"

"Five thousand years, at least," he said. His chin jutted slightly, suggesting that the trees would age another thousand years if I pressed him.

I asked Mahmoud to find me the head gardener, and he called over an old man in baggy pants who looked as if he loved his work. I repeated my question.

"They are very old," the gardener said.

"How very?"

"Very, very."

I mentioned that the guide said they were five thousand years old. He looked away and struggled to keep his gray mustache in a dignified straight line. Merriment lit his face.

"Maybe," he said. I rolled my eyes, and he laughed.

From the size of their roots and trunks, the trees had to be at least seven hundred years old. Botanists generally agree that olive trunks can't live much longer than that, but reasonable people insist that the trees between Jerusalem and Bethlehem go back to the Romans. The question is, how do you calculate the age of an olive tree? Even if you had a way to count the rings, it is not that simple.

An olive's massive root ball throws out shoots for centuries. When a main trunk eventually succumbs to insects, disease, and time, it can be sawn away and another will grow in its place. In the meantime, shoots at the edges of the root sprout fresh tendrils: new roots. Even if the old roots die, a tree still stands. The original stock is the same.

Hebrews planted that garden a very, very long time ago. Jesus sat under olive trees the night he was betrayed and Christianity was born. Muhammadans who conquered the Mount of Olives likely brought their fruit to Mecca. It may be possible to fix a date, but for me, that first impression was enough. Those twisted trunks had been there forever.

When I first pursued olives to Israel, in June 1994, Yasir Arafat was making his triumphal return. He came to set up autonomous rule after forty years in the wilderness. I went to Arafat's rally at Jericho, but he was late, and then very late. Crowds milled around behind a stout chain-link fence that guards had put up to protect their leader from the masses. But the masses numbered only in the hundreds. From the front it was a scene of distressing *déjà vu*. Palestinians pressed up against a wire, half fervid, half dejected. I went out to mingle.

One man in a kaffiyeh and clean shirt turned out to be Ahmed Nasser, a thirty-eight-year-old electrical engineer from Warren, Ohio, a U.S. citizen with family in Nablus, who had come to take part in the new Palestine. Like the rest, his political responses were guarded, but he was full of hope. "This does remind us of jail, and it hurts us a little," he said, gesturing at the high fence. A lot of the people there had been prisoners at one time or another. Nasser clicked amber worry beads and glanced around to see who might be monitoring his conversation with a reporter. Then I asked about olives, and he relaxed visibly.

"The olive has deep, deep roots in the ground," Nasser said, "and we feel that our roots are as deep. The Israelis know this. If anyone throws a stone from a field, they push all the trees off. We get punished for a thousand years." When Jewish settlers want to move into an area, he said, they uproot the olives that denote possession. Or they plant new trees to mark new ownership.

Every argument in Israel and the West Bank has at least six sides to it, and reporters are cautious about figures. But Nasser was right about the olive trees. The subject is so touchy that Palestinians near Nablus murdered a merchant who sold his old grove to Jews who wanted to settle there.

Many trees were uprooted during the *intifada* uprising, which calmed during the peace process that brought limited autonomy. But the Palestinians' Land and Water Office in East Jerusalem still keeps careful track of every bulldozer sweep and act of confiscation. From May 1993 to July 1995, it logged eighty-one entries, such as: "Cariut, 400 olive trees, settlers' aggression"; "Aboud, 60 olive trees, throwing stones"; "Jayyous, 400 olive trees, taken by Israeli quarry"; "Fara'oun, 15 olive trees, for security reasons"; "Beit Leed, 1,200 olive trees (reason not given)." The two-year total of 14,145 uprooted trees included figs, almonds, peaches, and others. Mostly, they were olives.

"Anytime you talk about Palestinians, you talk about olives," Nasser said. "They are in our culture, our diet, our way of life." He veered off into recipes and snatches of ancient wisdom. By then, others had formed a circle around us; they nodded in agreement or shouted derisively about matters of taste. Each had his own story of how Israeli administrators punished family and friends by bulldozing trees. They said it was impossible to sell their oil in Israel, or ship oil from Israeli ports, while Israeli olives were sold in the West Bank.

I remembered Chris Dickey's Olive Line theory: Beyond the Jordan, there is only desert, where people are without the civilizing force of ancient trees. That made sense in Jericho, at the eastern edge of the olive's range, where archaeologists have unearthed pits that date back to 8000 B.C.

When I finally had to break away, Nasser grabbed my arm. "Remember, there would be no Palestine without olives," he said gravely. "And without Palestine, there would be no olives."

Back at the American Colony Hotel, a fierce debate went on long into the night: Who made the best oil in the West Bank? Two Israeli friends argued hard for Jenin, far to the north through the timeless hills toward Galilee. Another pushed Nablus, a biblical city of labyrinthine lanes. Or Ramallah, a farming center nearer to Jerusalem. Finally, a fourth delivered an impassioned treatise on his secret source of oil. The next morning I headed for a village called Beit Ur El Fauqa, not far from Lod Airport.

On both sides of the road raw hillsides were dotted with huge dust-colored trees, full-crowned, with woody boughs heading off in every direction. In town I stopped at a little salmon-pink building with green doors. There was no sign, but sacks of cement and animal feed out front suggested it was the general store. A half dozen men sat in the sun, drinking tea and nattering. They all had olive trees. Naim Musleh, a thirty-nine-year-old merchant in a blue baseball cap, owned no land but tended five thousand trees for someone else. Mohammed Said Suliman, thirty, had seven hundred. Or eight hundred. He never counted.

An older man who gave no name, barefoot in a plum-colored polo shirt, offered a quick lesson in economics. He hired five people to harvest his trees. He paid each of them five kilos of olives a day—barely enough for a liter of oil—and kept the rest. If he did well, a month's picking produced two thousand liters, which sold for something over five thousand dollars. Deducting expenses, he might keep half of that. Sometimes his crop was extraordinary. In a bad year he was lucky to break even. This was no gold mine.

Their methods seemed *laissez-faire* and low-tech. "Some new varieties are being introduced around here," Naim explained, "but we like the old ways. Stronger trees, no sickness." The West Bank's standard tree is the malissi. Its small, gnarly olives make a wonderful, mildly peppery oil. Naim's friends pruned little, fertilized even less, and never watered. The approach seemed to have hardly evolved from the hunter-gatherer days.

Then I asked about grafting. A kid hurried off and came back with an armful of branches. Naim extracted a worn orange Swedish knife and flashed it around like a brain surgeon. With quick, delicate cuts, he made a set of barn doors on a woody stalk. Then he fitted in a smaller shoot, cut on an angle, just below the inner layer of bark. "Now you wrap this tightly with string," he said, doing it expertly. He showed me how to implant buds onto the end of a pruned branch, and then he demonstrated a half dozen other techniques.

These were Cato's tricks, as reported by Pliny, and improved upon by uncounted generations of olive people who redomesticated trees gone wild and improved the yield of slow producers. We did something similar in France, with store-bought wax to seal the graft. But whoever grafted that first oleaster, uncounted millennia ago, was most likely not more than a few days' donkey ride from here.

Naim had no agronomy degree, but he knew his olives. We had been talking through my friend Nihaya, a young Palestinian with a happy demeanor. Finally, we switched to what, for me, would be a lingua franca among olive people: pictures. I drew a tree in my notebook and, with pencil marks and sketches, asked about planting, pruning techniques, and pests.

The Palestinians' methods were much different from ours, but oddly similar. The secret to good olives anywhere is healthy trees. In Europe, groves are often pampered by someone with a lifetime to devote to them. But given a minimum of help, nature will usually manage. In either case, the basic principles are the same.

Olives appear only on two-year-old wood. In spring, a bud sprouts and grows into a long, slender sprig by winter. The cold weather "sets" the flowers; it triggers a process that results in the tiny, delicate, multi-petaled white blooms that cover these new branches in spring. Most of the flowers drop off in the wind and rain. Those that remain are hard little green olives by late summer. The young branches will produce buds for newer wood, but they will not flower again.

If you prune back wood that has given olives, your trees retain their shape. The olives stay low, easy to pick. The sap is concentrated and vigorous. If you let a tree grow, more new wood will sprout and you may have a larger crop. Eventually, however, you will need a helicopter to pick the olives. Some varieties, like those Palestinian trees, want to grow big. But they need pruning to control their height, remove cross branches that rub others in the wind, and eliminate old wood that has lost vitality.

Olives need air and light, but style is also part of it. French purists say a swallow should be able to fly through a tree, flapping its wings, without ruffling a feather. At least it must be visible from any angle if perched on an inside branch. The Beit Ur El Fauqa trees could hide a peacock in a Day-Glo T-shirt.

With the first warmth, the pests come. Growers worldwide have them in common. The worst is the dreaded *Dacus oleae*, the lacy-winged olive fly that deposits eggs into olives. This causes wormy-tasting oil, if the fruit does not drop off altogether. The *Coccus oleae*, black scale, clings to leaves and stems and deposits eggs by the thousand. Adults leave a sticky trail on the bark, which attracts a fungus, the familiar *fumagine*. Left untreated, the bark looks as if it has been charred and limbs die. There are borers, moths, and spiders, not to speak of tuberculosis tumors and other diseases. Nature needs help to alter the balance in favor of healthy olives. Copper compounds and poisons work, if used correctly. Otherwise, they tilt the balance in another direction.

In spring, growers dig around their trees to the width of the crown. This aerates the soil, helps water penetrate, and removes competing plants. Those who can afford it add nitrogen, in one way or another—it helps if you have sheep. Year-round, it is open season on the shoots which grow by the score around the trunk base, as well as suckers on the lower limbs.

And then there is moisture. Early in spring, you pray for rain. When the flowers come out, you pray that it stops. A late-summer soaking is most welcome provided it is not hard enough to knock olives off the trees. Autumn hail is deadly to olives and branches alike.

Naim nodded with vigor as I ticked off the various problems. We compared notes on solutions. He roared at my feeble attempt to sketch dacus wings.

Finally, I asked about harvesting. "We always pick by hand," he said. "Otherwise you damage the tree for the next year." This time I nodded with vigor.

Palestinians, like the Israelis and the French, are milkers. Many Spaniards, Italians, and Greeks are whackers. It is a question of economy. Milkers run their fingers quickly down a branch, detaching as they go along. They put the olives into a basket or drop them onto a net under the tree. Whackers use a stick to hit the branches, knocking off olives and a good bit of tree at the same time. This is much faster and allows crews to harvest tall trees without a ladder. It also breaks off the tips of new wood needed to grow more olives.

In a good year Naim and his friends might pick 150 kilos on a large tree, three to four times a respectable yield in Europe. That means months in the air on a precarious ladder. But for all the work, he was a happy grower.

"We have altitude, and mountain olives give better oil than those from the plain," he told me. This is generally true because of rockier soils, fewer insects, and the mix of morning dew with dry afternoons. "Our olives are better than those in Nablus or Ramallah. And our oil is the second best in Palestine. The only better is at Beit Jallah."

Where, I asked my driver discreetly, is Beit Jallah? Within minutes we were hurtling toward Bethlehem. First, however, we lingered briefly for departure ceremonies. Hands were shaken firmly around the circle. The last cups of sweet tea were drained. Naim bowed solemnly and said, "May Allah make your olives grow and grow." Why not? It worked for him.

I found Farid Mukarkar atop his mountain above Beit Jallah, just past the ramshackle Hotel–Café Everest. He emerged from his sturdy henhouse looking puzzled, tugging at his sagging waistband with his left hand, and dribbling sick-chicken goo from a syringe with his right. His wife, a solid woman, brushed black hair from her face and eyed me coolly. A Jewish-looking stranger arriving unannounced was not necessarily a good start to the afternoon. When Nihaya explained in Arabic who I was and what I wanted, Farid beamed a sunny smile, and dropping the needle, he clasped my left hand in both of his.

Farid had a friendly round face, with a bristling gray mustache and a fringe of white hair around a sun-burnished bald head. His eyes ranged from shrewd to elfin, and he always seemed to be chuckling at a joke he forgot to tell.

Nihaya mentioned I was an American; Farid pointed at me and roared with mirth. "Ha!" he said. "Football!" Apparently the United States had just been humbled by Equatorial Guinea, or Grenada, in the World Cup soccer tournament. When she narrowed my origin to Arizona, he slapped his pockets in a mock fast-draw. This was definitely my guy.

We had found Farid by asking around. Beit Jallah is a stone village near Bethlehem, settled by Christian Arabs at some hazy date during the time of the New Testament. Its life went on in hidden courtyards and murky cellars behind shuttered windows. Yes, people told us, Beit Jallah produces the best oil in Palestine, and someone pointed us up the hill to Mukarkar's place.

"Lots," Farid replied when I asked him how many trees he had. This, I learned later, meant five thousand. I asked how long they had been in his family, and he gave a helpless shrug. "I took the land from my father, who got it from his father, who got it from my great-grandfather . . ." He went on adding greats for a while. "Well, we had it during the Turks, and that lasted four hundred years. And then before that . . ." He threw his hands up.

His trees stepped down a hill terraced by thirty stone walls, flanked by spreading apricots and apples. They were massive, long-branched, and thick as the trees at Beit Ur El Fauqa, but expertly pruned and full of fresh wood. Mostly malissis, the good ones produced eighty to ninety kilos. "I cut them back every two years," Farid said. Scissoring his fingers at my shaggy neck, he added, "Not much. Like a haircut."

It had not been a good year. "The weather is changing," he said. "Winter used to be January to February, and then we'd get a good soaking. That is not so often anymore. This year the trees got no water in January or February. Everybody waited for rain in March. Nothing; so there was no decent fruit. We need sporadic rains from January to May. The trees need more water, but we have none. Israel gets the water. We need help to get rid of the olive fly—chemicals—but we get nothing."

Later, from the terrace of Farid's stone house, we watched Israeli progress at work. Far down below, bulldozers had cut a great gash into the hill, digging a tunnel as part of a highway that would link Egypt to Syria, the new Middle East. Right-of-way land is confiscated with no payment, he said. "They destroy many, many old trees. They offer to move them—in the middle of summer. They come to us and say, 'Where do you want us to put it?' " He threw up his hands. "What difference? It will die."

Farid talked politics with sadness rather than passion. A Christian, he had no religious problems. A farmer with his own little family empire, he had no further ambitions. He just wanted to grow his olives and vegetables and raise chickens. He was not particularly thrilled by Arafat's return and preferred to wait and see what it meant in the long term. Meantime, he knew that constant conflict in the West Bank meant collective punishment.

"For years we drove to Jerusalem to buy what we needed, to live our lives," he said. "Suddenly that stopped. We are cut off. How do we survive?"

The mood lightened decidedly when we shifted back to olives. Farid and his wife, Madeleine, were amused at my claims to be an amateur oliveman.

"After you were born, did your mother wash you in water and salt and then rub you with olive oil?" he asked. I wasn't paying attention, I said. It was probably chicken fat.

"How do you keep your oil?" A new test.

"Ummm," I non-answered.

"My way is better." Matter-of-fact, not boasting. "We use pottery crocks, in the dark. This takes all the acids from the oil." I had seen them, sixty-liter porous containers with a bulbous bottom and a narrow neck. "If you have it fifty years in the same jar, nothing will happen."

Actually, the old pots were glazed inside. Porous clay did not change the oil's chemistry; that altered itself, over time. And when the level drops in unglazed pots, deposits on the sides tend to get rancid, ruining the good oil. Fifty years was definitely pushing it. But I was not about to argue olives with Farid Mukarkar.

"Look at me," he said, whacking his bicep. "Sixty-seven, strong as a bull. I live on olives. I take a piece of bread, dip it in oil, sprinkle it with thyme, and eat it. Why all this steak? My uncle is 110 years old. If you fight him, he'll beat you up. He rubs himself every day with oil. He eats olives. He has worked with them all his life."

Someone had told me Palestinians sang special songs when they picked olives.

"They did," Madeleine said, a little wistfully. "Not anymore. There used to be a real spirit. When we started, there were no trucks, only donkeys. Fifteen men carried the ladders. Women brought sheets to spread under the trees. The older women chanted and wailed. Now it's only the radio. There's no more spirit to sing and have fun."

She remembered a few old Arabic verses. One went: "Olive, my aches and pains are your fault; I'm going to pluck you, squeeze the oil from your eye and take you home." Another song was a prayer that the olives would turn to lemons so they'd be easier to pick.

Lunch aromas wafted from inside, and I asked Madeleine for her favorite recipe with olive oil. It was *jaz mussakhkhan*, a twice-cooked chicken mashed onto tasty, moist pita flat bread. Like most Mediterranean cooks, she expressed measurements in terms of cupped hands, rubbing fingers, and a shrug. She cuts a young chicken in two and boils it to make broth. She marinates the boiled chicken in lemon juice, sumac, and other spices. Then she sautés it in olive oil and chopped onions. Finally, she moistens flat bread with the broth, adds more sumac, presses the chicken on top of it, and bakes it.

I asked Madeleine how she cured olives for eating. Her method was identical to Naim's, in Beit Ur El Fauqa. It was also how Nihaya's family did it. First, they crack ripe green olives slightly. The children used to do this with stones; now a specially designed crank-operated grinder does it faster. A wooden mallet also works. Soak the olives for a few days in plain water, changing it two or three times. Then fill a large container a third full of fresh water. For ten pounds of olives, mix in about a half pound of salt, a half pound of hot green peppers, and some lemon. Cracked black olives are better with red pepper and no lemon. There's a trick to testing the amount of salt, which is crucial. The water should be just salty enough to float a raw egg. After ten days, the olives are good: crisp and tart. After a few weeks, they're great: chewy and sweeter.

The old people insist on using well water. Some add garlic. A few use vinegar. If they want olives to last a few years, they don't crack them first. The oil inside wards off bacteria.

Green olives take more curing for the same reason that early-season oil is sharp. The olive fruit's main constituents are water, sugars, proteins, anthocyanins (pigments), and oleuropein. This last compound—a phenolic glycoside—is healthful but bitter. It diminishes as olives ripen and synthesize more oil, and it is soluble in water.

I had a plane to catch and had yet to taste any oil. With no time for subtle hints, I tried a frontal assault. "Um, Farid, do you have any oil to sell?" I asked. "No," he said, "it was all sold right after pressing." With a sly glance at Madeleine, he added: "But maybe we can find some to give you." She hurried from the room. Moments later she emerged with a Johnnie Walker bottle stoppered with a cork. The oil was a deep green gold, thick but clear, filtered by natural sedimentation in an underground clay crock. It was all I could do to keep from splashing some on my fingers to taste it.

"Gold," Farid said, holding it up for a last loving look and then wrapping the bottle in a newspaper. I burbled thanks and promises to bring my own oil on the next trip. I'd come back to help him pick in the fall. "Of course you will," he said, with the air of a man who has heard a lot of empty promises.

By then, I was late. With any sense I would have switched to an Israeli taxi with yellow license plates for a fast ride to the airport. But my Arab driver had done well for me. I asked if his blue West Bank plates would delay us at checkpoints. "No problem," he said. He took a slow back road, avoiding the fast Jerusalem–Tel Aviv freeway. "No problem," he repeated, as tiny beads of sweat appeared on his forehead. The first checkpoint cost us ten minutes. Finally, with seconds to spare, he announced with a toothy grin, "Airport."

At the outside gate a nasty young soldier motioned us over behind four cars and two vans. Guards poked into every bundle and crevice. The line did not move. Israeli cabs whizzed past without slowing down. I tried to reason with the officer in charge. He asked why I was in an Arab cab, and I told him. "Olives?" he snorted, directing me back to wait in line. Finally, we got through, and the driver screeched up to the curb.

Halfway out of the taxi, I sensed something was wrong. Taking two more steps, I realized what it was. The soldiers had searched my pack, and

I'd neglected to strap it closed. I felt the weight shift and made a mad grab. It was too late. Farid's bottle of gold smashed on the concrete.

In this land of clashing symbols, I assumed that someone was trying to tell me something. But, after five millennia of to-ing and fro-ing, who could tell what it was? I restrained myself from plunging a finger into the shattered glass, on the filthy walkway, for a tiny sample of Beit Jallah oil. Instead, I decided that nothing on earth would keep me from picking when fall came.

That October I called Nihaya to find out when Farid would be harvesting his olives. "He is almost done," she reported back, "but he'll save you a tree."

This was startling news. My own olives were still hard green stones. Even allowing for the climate differences, his harvest seemed awfully early. When I got there, Farid proved true to his word. His crop was at the mill, but my tree was waiting. It was a monster.

"Okay, let's go," Farid said. He handed me a heavy wooden ladder that was four times our height. I tried to drag it with little success. "No, no," he said. "Upright." With both hands, I struggled to keep the thing vertical, while he held his sides and shook with amusement.

"Like this," he said, grabbing a low rung, tucking the bottom part under his arm like a cheerleader's baton, and trotting off with seventeen feet of ladder balanced above his head. Then he did a little dance.

Farid nestled the ladder in among the branches, against a sturdy tall limb. I scampered up it and began to pick like a demon. Farid waited for another good laugh. But I'm a good picker; fruit flew down. No leaves, no rejects, just a steady hail of olives that were ripe as they could be for early October.

A chilly drizzle turned to hard rain, and I kept picking. So did Farid. He looked up slyly a few times to see if he should call a halt. "You don't really have to do this," he said, assuming from the start that I had not planned on doing any real work. "Hey, I eat olives," I said. "I'm tough." Soon we had a fair pile, and I was getting into it. After the first few minutes, olive picking is not much fun. But competent milking takes enough skill to offer some satisfaction.

Climbing almost to the top of the ladder, I dived into fresh territory. By way of conversation, I called down: "Does anyone ever fall off these things?" No sooner had I said it than the ladder began to slip. I jerked sideways to keep my balance, and it dropped from under my feet. Grabbing wildly, I caught a branch of questionable strength. It bent under my weight. Gingerly I got my left foot onto something solid. Like an inept orangutan, I worked my way to the ground. Farid guffawed. "Dream you'll see a monkey," he said, "and you'll see a monkey."

We stopped and caught up on developments since my last visit. Things were going well, he said, but his 110-year-old uncle had died. Olives were not magic. Farid's extended family had helped in the harvest. Altogether, there are three hundred Mukarkars and five thousand olive trees. His sons dabbled in other careers, but all would inherit land. Ibrahim, educated in Sweden, ran a factory that made elastic for underwear. But he still tended the trees. Issam, the youngest, was going to marry at twenty-two. He studied business but wanted to stay close to olives all his life.

Farid took me to the mill in Beit Jallah. Shaky English letters proclaimed its name: The Cooperative Society for Pressing Olives and Marketing Its Products—and the Modern Cooperative Soap Factory. The ramshackle buildings looked as if they housed something ancient. Inside, two modern Italian centrifugal units churned out oil at a rapid rate. There was no sign of soap.

I had expected something like our timeless mill at Tourtour, with old-fashioned mat presses and open tiled decantation vats. This was not the style in the West Bank. The old way took too much work and too much time. The wealthy Beit Jallah cooperative had already advanced several generations. It had abandoned its first centrifugal mill, a post-industrial mechanical monstrosity stamped AIX-EN-PROVENCE with no manufacturer's name. Instead, it used a Rapanelli continuous system, which could mill four tons in an hour. Sentiment was not a strong selling point. Palestinians hauled in olives and, not long after, hauled out oil.

The process was simple, one I would see over and over again throughout olive country. By truckload or by old burlap sack, olives are weighed to determine each producer's share of oil. Then they are dumped in a hopper. Once washed, they ride a ribbed rubber conveyor belt into stainless-steel horizontal centrifuges. Grinders spin the olives into mash.

Water heated above 85 degrees Fahrenheit is piped in to liquify the paste and help separate the oil from the olives' vegetable water.

Oil is extracted by breaking down cell membranes in the olive pulp, whether by pressing or spinning. Hot water helps this separation, but it destabilizes the oil. If water is used—and purists shun it—the trick is to get the right temperature. Any added water can wash out volatile flavor compounds, particularly if it is hotter than 80 degrees.

From the horizontal units, mash is pushed through conduits into the vertical centrifuges. These whirl at high speed. The water, black and greasy, spins off to flow out of one pipe. The finished oil spills into side-by-side gleaming steel tanks.

But the scene at the mill was as old as the hills around it. Men slung their sacks onto the scales and went off to see their pals, swaggering slightly if their production was noteworthy. Kids played all around, happy to be done with picking. Women in shawls squatted by the tanks with rows of jugs, waiting.

When the oil is done, families draw off their treasure from a spigot at the bottom. On one tank someone had taped a large sign: MAY ALLAH BLESS THIS OIL. They keep a year's supply for their kitchens. They sell the rest in markets or at home, by the kilo, in chipped jugs, yellow jerry cans, or plastic water bottles. Depending on the year, they charge something close to five dollars a liter, after paying a fee to the mill. If they count nothing for all of their hoeing, pruning, picking, and hauling, that is mostly profit.

Elias Jahshan was proud of the plant he had recently come to administer. At forty-seven, he was starting a new career. "I used to work for Israeli customs during the *intifada*, but I quit," he said. "This is better, a much more satisfying life. I like the idea of serving. Here we crush olives. There they crush people."

If the crop is good, Jahshan runs the mill twenty-four hours a day, from late October until the new year, if necessary. Some seasons are awful, he said, but there are plenty of trees. The old ones, known as Roman trees, are getting scarcer. Under the Turks, every eleventh tree was cut to fuel the trains. Some were sacrificed for urban expansion or settlements. But fast-growing exotics are planted all the time, by settlers, or Palestinians, to mark territory. An olive tree is nine-tenths of possession.

At the Beit Jallah mill, the oil's acidity ranges near .4 percent, well under the 1 percent standard for extra virgin. "It is very good," Jahshan said, "but that is a matter of taste. Many Palestinians like it strong, 2.5 percent, even up to 4 percent, so harsh it burns their throat on the way down."

A friend of his had taken some Beit Jallah oil to his old mother in Nablus for tasting. She spat it out and gagged. "Water!" she pronounced with disgust.

Farid's fresh oil had a fierce kick to it. He shrugged. "It will get better later. Last year, we had to let the oil sit for six to seven months before it reached its best flavor." That is what happens when you pick and press early, before the ripening process reduces the content of oleuropein. With time, the chemical balance shifts and the oil mellows. But early oil is early oil, a taste some people love and others never acquire.

In earlier years, Farid had let his olives sit around in sacks while collecting more for a large mill run. Some ripened too much and spoiled. Acidity soared, and the oil tasted bad. The previous season, when he experimented by taking the olives straight to the mill, the oil was too sharp. This time, he aged most of his olives by spreading them in the sun on his roof.

The question of acidity is complex. Olive oil is graded by the number of grams per hundred of free oleic acid, but that is only a means of measurement. When olives are left unpressed, fermentation drives up their acidity. But other components, not oleic acid, produce the harshness and determine the taste. An oil of high acidity can be flavorful and mellow.

Bitterness in fresh oil goes away not only because oleuropein dissolves but also because elusive polyphenols—antioxidants—break down with time. As a rule of thumb, the fresher the oil, the more healthful it is.

Since the ancient Greeks, olive people have determined the flavor of their oil by choosing the right time for picking. I asked Farid why he didn't wait longer to harvest, a question I repeated all over the West Bank. The answer was not agricultural but sociopolitical. During the turmoil that followed the creation of Israel in 1948, a lot of Palestinians stole olives from their neighbors' trees, especially where ownership was vague. Cruising Bedouin families could devastate a grove overnight. The only safe place for olives was in jars or oil. Families picked as soon as nature allowed.

Now olive security is less of a strategic issue, but the habit remains. As soon as one family in the neighborhood starts to pick, the others follow. No one wants to be stuck with unsold oil in case the price peaks and drops.

Back at the farm, Madeleine made us lunch, a heaping plate of rice topped with chicken fried in olive oil. We splashed Farid's sharp, fruity oil on a salad and dipped bread in what was left on the plate.

Farid badgered me mercilessly about my divorced status and, particularly, my lack of children. Along with five sons, he and Madeleine had five daughters. "I'm going to get married to another woman and have ten more children, and you still won't be married," he said, and I laughed along with him. He was safe as long as he didn't bad-mouth my olives.

Over strong coffee, Farid suddenly stopped joking. So many questions about the present and future had a sobering effect.

"Twenty years ago the trees were bigger," he said. "The soil has nothing new to give. Every year, smaller and smaller. No more vitamins. There is no rain. Whether you try chemical fertilizer or manure, it is the same. You need rain. Most of the trees are sick. Look, human beings face new diseases every year. Sometimes there is no solution. No cure. Mainly, we don't get enough rain. The weather is changing. And the Israelis give Palestinians 2 percent of the water. The rest goes to settlements and Jerusalem."

This, he went on, meant a social and economic change. "The young people are going to factory jobs, other things. They cannot live on the land. When people get more and more numerous, do you think the size of each country will get bigger and bigger?

"I have a hundred dunums of land [twenty-five acres], and I must divide it among my five sons," Farid said. According to local custom, women seldom inherit land, which might end up in their husband's hands. "They will have more kids, who will need land to build on, agricultural land. Year after year we will lose these trees. I see it happening with other families."

I asked Ibrahim, the Swedish-educated son, if he agreed with his father. He nodded, not happy. "I don't think we're going to have olive trees very much longer," he said. In Europe, he added, governments help people nourish the land and protect it. In Palestine, the land is simply used.

Sitting atop his hill at the edge of Bethlehem, looking down over trees that might have been there since Christ was born, Farid Mukarkar grew yet more melancholy. "Everyone says the same, whether Muslim or Christian," he said. "This looks like the end of life. Everyone is killing each other. Death is everywhere. We are not getting anything out of the soil. This is the end of life."

Then the mood changed. Farid talked about my next visit. He used no conditional verbs about the presence of olive trees, just a straightforward future tense. They had survived thousands of years on his mountain. They would not give up easily.

One thing was still on my mind, and it took the subtlest of hints. Madeleine disappeared and came back with another whiskey bottle. It was the good stuff from the year before.

"Gold," said Farid, his old self again, holding it up to the light.

Nothing was going to happen to that bottle.

At the Beit Jallah cooperative I'd asked Jahshan what happened to the gritty gray-brown mounds that collected each day after milling. When oil is extracted, waste known as pomace is left behind. Usually it is sold to refineries, which use solvents and high temperatures to leach out inferior oil. In the West Bank, merchants from Nablus buy up all they can. The craft of making Nablus olive-oil soap is as old as dirt. In the stone warrens of that ancient little city, Jahshan told me, families still do it the old way. The next morning I was off to Nablus.

This, at the time, was not the wisest of ideas. Violent encounters between Israelis and Palestinians had the West Bank on the boil, especially Nablus. Muslim extremists were looking for incidents to create. Young Arabs expected to find Israeli agents in something besides a uniform. But Nihaya was pretty crafty, and in any case, a faint heart never won a bar of Nabulsi soap.

The morning started well. My Arab driver declaimed on his own preferences. "I like Spain oil, very low acid," he said. Each year he bought bulk amounts for labane, hummus, and eggs. On the way, we stopped so I could pick with a family by the side of the road. The old man appreciated the help. We discussed diseases and pruning like a couple of old olive bores.

"I was picking olives before I could walk," he said. "I'll be picking them when I die."

When we arrived in Nablus, it was pounding rain. Nihaya and I were joined by a friend who'd come along for the fun of it. A dozen sharp turns later, pushing through crowded, cobbled streets wide enough only for a laden donkey, we came to a river. Water rushed down a cross street, deep enough to float Noah's Ark. Nihaya, who is very short, waded across without a pause. She was soaked to her hip pockets. I started to follow. "Unh-unh," my friend said. "I'll meet you back at the car."

I could not send my friend, Jewish-looking and draped in black Leicas, back alone. The mayor of Nablus would not have found his way without help. Nor could I send Nihaya. I'd be in the same predicament, with no translator if I found the soap. Just as I was wondering what I'd tell his trusting wife if I came home without him, a storm drain unclogged. The waters parted.

We almost missed Mohammed Majid Nabulsi's factory, with its small carved stone doorway. We passed through and tumbled down ten centuries. Grimy characters out of some ancient book were dragging a giant swizzle stick through a bubbling vat. Dim lights like oil torches barely illuminated the vaulted stone ceilings. Nabulsi, sixty-two and spry, hopped up from his account books and came over to show us around.

The process is simple. Olive waste is boiled with caustic soda. A thick layer of oily scum floats to the top and is scooped up in buckets. Only then does modern technology kick in. The buckets are placed on a small platform and hauled to the upper floor on squeaky wheels, along a rail set into the stone stairway. The motor sounded two decades late for its annual overhaul. We took the steep stairs, which were coated in hardened slime.

Upstairs, workmen slopped the molten soap into frames of timbers laid in large rectangles on the stone. It was exactly like pouring a concrete floor. In two or three days, after it congealed, a soapmaker dragged a long razor blade on a stick across the stuff, cutting perfect cubes. These were wrapped in paper bearing handsome Arabic script and shipped. The factory turned out thirty thousand bars each olive season.

Nabulsi made two kinds of soap. The green bars, for washing clothes, were made from the remains of the West Bank olives. The better kind, pure white, started as coarsely refined oil. This was imported from Italy for the equivalent of 35 cents a kilo. Local oil would cost him $1.50.

"We eat almost all the oil we produce," he explained. So do the Italians. The Italian exporters had probably bought the oil somewhere else, part of a Byzantine international trade pattern that I'd explore later. "In the past, what was left behind contained enough oil to make good soap," Nabulsi said. "Now there is better machinery to extract the oil, and all of it goes for eating."

He could not resist a friendly dig at a visitor from extra-virgin land. "Here we eat oil. You eat water." He was less flattering about foreign soap. "Olive oil is natural, cleans so well. But if you see soapy flakes, foam, you know you have something else. That other stuff"—he named a few brands, including one with olive in its name—"is 80 percent animal fat, with other oils, perfumes. No olive."

Nabulsi's family had acquired the factory in 1880. He does not know when it was built or who used to run it. Back then, perhaps forty families made soap in the Nablus casbah. Now there are four or five. Most of the soap comes from modern plants at the edge of town. Nabulsi made up little gift packages and offered tea. Then he sat down to his water pipe. We went to find a restaurant that cooked with oil which did not taste like water.

Israelis also love their olives. The same engineers who grow succulent oranges in sand are testing the limits of the ancient olive. The Volcanic Institute, a research center not far from Tel Aviv, has been probing into the olive's secrets for forty years. An experimental station at Sde Boker, David Ben-Gurion's old kibbutz deep in the Negev, breeds olives the size of golf balls. Pithy and full of more water than nature intended, they are not taste treats.

In Israel, good olives are everywhere, on breakfast tables, on store shelves, in kitchens. One Sunday morning I pulled off the Jerusalem–Tel Aviv freeway into a gas station. It was closed. The only oil around was in small bottles at a snack bar out back, green-gold extra virgin from a kibbutz called Tzora, with roots in the Bible. Had it been a Saturday, I'd have been happy with oil from the Trappists' domed Lathun monastery on the hill.

I drove over to the monastery and through an iron gate, which opened into an earthly paradise: lush gardens, moss on old stone, shade trees. A young French monk looked over and shot me as evil an eye as a professional Christian is allowed to give a visitor on a holy day. He softened slightly when I said I only wanted to know about olives. This was not his area, he said, nodding slightly at a graying monk whose area it obviously was. The old monk approached us, but it was still Sunday. He collected his younger brother, and they moved off without another word.

The exit road wound through the orchards, so I parked and took a walk. It was pruning time; foliage was piled high among the trees. Each cut had been carefully made. Thick cross limbs were dispassionately sawn away to make room for the new. In Europe, we make drastic cuts only after the winter freeze. In the warmer Holy Land, they lop off laden limbs and pick at ground level. A breeze ruffled the leaves. Graceful boughs dipped under blue sky. I lay down against an old gray bole and breathed pure air, peaceful among the olive branches.

A few months later, not many miles away, Jews and Arabs were warring over olives. Jewish settlers wanted to expand their community, Efrat, near Al Khader. To lay claim to disputed pieces of land, Palestinians planted olive trees. Settlers tried to uproot the saplings. Villagers tried to uproot the settlers. Israeli border guards rushed in by the dozen, beating the Palestinians with rifle butts and fists. Saeb Erekat, Arafat's minister of local government in the self-rule authority, was knocked out and dragged away.

"This is the graveyard of the peace process," Erekat pronounced, when he regained consciousness.

It wasn't, of course. The peace process has been operating, interspersed with war, since before David slung his rock at Goliath. After I left, a Jew murdered Yitzak Rabin, the beloved prime minister who had shaken Arafat's hand. Much of the West Bank passed under Palestinian rule, including Nablus and Beit Jallah. Soon what had been unthinkable was commonplace: blood enemies went into business together. One of the first enterprises was only logical. A Jew from Jerusalem and a Palestinian from Gaza, together, began exporting olives.

———

Heading south, I drove back in time three thousand years and found a historical surprise. The average Philistine, by reputation, is not a likely candidate to review art for *The New Yorker*. In fact, the Philistines were a cultured nation of sailors and artisans who got rich with the first Middle East oil boom. Three thousand years ago, they planted huge olive groves and built hundreds of stone presses along the coast near Gaza.

The Philistines came from the Aegean in the twelfth century B.C., and possibly from Crete. They fought early Hebrew tribes; Goliath was one of their larger commandos. The treacherous Delilah was a Philistine, and it was their temple that Samson destroyed. Their name is the basis for the word "Palestine," a geographical term the present occupants prefer to sub-stitute with Holy Land. Much more is now known about the Philistines, thanks to a gentle-mannered Israeli with thinning hair, Natan Eidlin.

I found Eidlin in his fascinating little museum at Revadim kibbutz near the digs at Tel Miqne, not far from Ashdod. Lanky, with a slight sloop, he was a kindly Jewish uncle, scholarly but tough and sunburned. A teacher turned archaeologist, Eidlin spent many of his early years poking around ancient rocks. He was born in Haifa and went in 1948 to help found a kibbutz in the Hebron hills. While he was off fighting the war, Jordanian troops overran the kibbutz and imprisoned its residents. Eidlin had met a girl from Revadim, and he settled there.

Walking around one day, he stumbled onto an intriguing tell, a mound hiding ancient remains. He found some skilled helpers and began to dig around. In one corner, he saw the traces of a familiar structure. "I knew it was an oil press," he said. "I had seen such a press thirty years earlier in Galilee. It was operated by a screw and a donkey. I told the others, 'If we are lucky, we'll find the crushing vat.' " Soon, they had found crushing vats by the dozen, along with hundred-pound granite hand rollers that were used to smash the olives.

Eidlin pored over biblical passages and also pinpointed archaeological sites that were already identified. He worked out trade routes and logical communication roads. From old accounts, he studied how armies moved in battle.

Eidlin studied the biblical account of how the Philistines escaped with the Ark, hauling it by oxcart. He figured out the route that oxen would have had to take, over hills and along the wadis, riverbeds. He displayed such faith that I asked if he thought the Bible was the unfiltered word of God. "I don't believe the Bible came from heaven. Somebody wrote it," he said. "But the guy who wrote it certainly knew the territory." Finally, he determined that he had discovered the Philistine capital of Ekron.

The Israelites hated Ekron. When the Philistines slew King Saul and stole the Ark of the Covenant, with the first-edition stone Ten Commandments, they took it to their capital. King David sacked Ekron early in the tenth century B.C., but with its oil wealth, the city-state rose again. It allied itself with the Egyptian pharaohs, who offered little help when, in 603 B.C., King Nebuchadrezzar of Babylon leveled Ekron and herded the Philistines off to slavery.

In 1980, Seymour Gitin of the William F. Albright Institute of Archaeological Research and Trude Dothan, a specialist on Philistines at Hebrew University, organized an American–Israeli expedition. They dug out part of Ekron and traced grand lines. The artifacts they found were the pride of the Iron Age: red and black ceramics painted with birds and fishes; an ivory handle carved with a woman's smiling face; bronze juglets for olive oil; beer mugs and tableware. Later, a stray piece of wire in the dirt turned out to be a golden headband in the shape of a cobra, once part of the statuette of an Egyptian goddess.

Gitin and Dothan moved on to other projects. Eidlin remained in his kibbutz, running the museum. He will talk for hours with very little prompting. His impeccably researched patter, delivered in soft-spoken, witty singsong, might be a monologue from Yiddish theater.

"Olive oil was among the most important products in the world at the time," he said. "It shaped cultures, provided jobs. There were daily uses, holy aspects, the rites of kings. Large pottery industries grew around it. If a prophet wanted to utter a curse, he would say, 'Let God ruin your olive trees.' "

Eidlin had spent years tracing the olive's origins. In Jericho, he said, archaeologists have found the charcoal of olive pits dating back ten thousand years. Almost every dig in the Near East suggests people depended heavily on oil.

By the time Ekron began exporting large quantities, olive oil was far more important to daily life than petroleum is today. It was the only source of light. People cooked with it. They made cosmetics and medicines from it. Status was determined by who could afford to have themselves oiled down regularly. Oil was essential for temple sacrifices. Great seagoers, the Philistines carried their oil to Egypt and around the eastern Mediterranean in simple open vessels. In 712, the Assyrians seized Ekron. They expanded oil production and found new markets.

Eidlin took me to his chef d'oeuvre in a nearby grassy park. He had re-created an Ekron press, using original crushing vats and stones from the dig. It looked a lot like the old mills in France. Olives were crushed under granite cylinders, like huge rolling pins, which millers pushed back and forth for about twenty minutes. The mash was placed in round mats woven from tough grass, one atop the other, and these were stacked on the hardwood frame of a press. Instead of the massive screws that old-style presses use today, the Ekron millers applied pressure by leverage. They tied four heavy stones to its long wood beam.

Oil was separated from water in large clay jars with holes along the side plugged by carved wood stoppers. Sometimes hot water was added to hasten the process. After a day or so, the oil rose to the top. The bungs were removed to drain the water. Bronze dippers lifted out the oil, leaving the last sediment and water at the bottom.

Eidlin estimated that Ekron's 114 large mills each produced seven to ten tons of oil a season, working four hours a day for three months. That comes to a thousand tons. If the population numbered at most seven thousand, there was a lot to sell.

No olive trees remain. Many disappeared when the city was destroyed. Back then, farmers planted wheat among the rows of trees. One attacking army sent three hundred foxes into the fields, with torches on their tails, and left a lot of charred stumps. Time and weather did the rest. Even with the surviving artifacts and artful mock-ups, it is hard to imagine those lush groves that once covered the dry, dusty steppes and scrubland that stretch southward to the empty Negev.

Eidlin has plans to expand the setting, to plant a new grove of olives and re-create an ancient mill, using as many original pieces as possible. He wants schoolchildren and their parents to appreciate the roots of civilization which are slipping away in an age of technology. "One day," he said, "I hope you will see a park here, with trees all around, where people can come and press oil in the old way."

Spanish Picual

Fresh Tuna Escabeche

María José Sevilla, whose passion for cooking started when she toddled around her mother's kitchen, laughs when asked for her favorite olive-oil recipe. "That's a foreigner's question," she says. "In Spain, everything is cooked in olive oil." But she likes escabeche made from the small white albacore tuna.

1½ pounds fresh tuna, in one piece	2 cups extra-virgin olive oil
Salt	2 bay leaves
10 ounces onions, peeled and chopped	2 sprigs of thyme
1 small head of garlic, cloves peeled and sliced	7 ounces sherry vinegar, or white wine vinegar
	¾ bottle of dry white wine

Season the tuna in salt and set aside for 2 hours. Fry the onions and garlic in the oil, adding the bay leaves and thyme, until the onions are tender and translucent, not brown. Place this with the tuna in a deep flameproof casserole, pour in the vinegar and wine, and add enough water to cover. Season to taste with salt. Bring to a boil, then reduce the heat and simmer until the tuna is very tender, about 45 minutes. Remove from the heat, leaving in the liquid, and refrigerate. Marinate at least 24 hours before taking the tuna from the stock. Serve with caramelized peppers, salad leaves, roasted vegetables, and some of the stock.

Serves 6 as appetizer.

Adapted from *Mediterranean Flavours*

4

Olive Heaven

In the heart of Andalusia, Paco and Andrés Núñez de Prado offer elaborate apologies for everything in their pressing process that has changed since the Philistine presses at Ekron. There is not much to excuse. Instead of hand-powered stone rolling pins, their cone-shaped granite crushers are driven by some late Iron Age machinery. The tower of mats is squeezed not by stones hung on a wooden lever but rather by a 1936 hydraulic motor that works so slowly you know it is operating only by its wheezing cough.

The brothers' oil bears the treasured shield of Baena, one of four Denominación de Origen (DO) labels recognized in Spain. Their fanatical purism is not the reason. Large local cooperatives also qualify, and they use high-speed centrifugal continuous systems with computerized digital display control panels that look like the dashboard of the starship *Enterprise*. It is simply that the Núñez de Prado brothers are passionately, desperately in love with olives and everything about them.

I made a pilgrimage to Baena in November, just as the olives were turning black and the lovely fall colors faded into a stark beauty. Andalusia, Spain's largest and most populous region, stretches from the Mediterranean and the Atlantic Ocean to the steppes of Don Quixote's La Mancha. Phoenicians settled it a thousand years before Christ. The Romans took it from Carthage and made it their rich province of Baetica. The Moors came in 711, leaving behind exotic, crenelated forts after Castillians drove them off nearly eight centuries later.

I saw Baena long before I reached it, a small white town with a cut-stone skyline perched upon a hilltop. The Núñez de Prado brothers I found by a cheery olive-wood fire in the brick hearth of their mahogany paneled office. Their family coat of arms hung on a wall among *objets* dating back to Isabela la Católica. There was a monster iron safe. A gigantic carved table, littered with fountain pens and gold-stamped ledgers, stood by overstuffed, cracked leather armchairs. The lovingly maintained Smith Premier typewriter was so old it used two keyboards: one for lower-case letters and another for capitals. Only a muttering fax machine suggested the 1990s.

Paco had studied diplomacy and spent three months in the Spanish Foreign Service in Vienna before coming home again. He wears baggy English tweeds and Italian silk ties. Andrés, an agronomy engineer, never took his eyes off the trees. He wears baggy English tweeds and an ascot.

Andrés, the shy one, looks after the technical end. Paco, with slicked-back hair and the slightly crazed gleam of a likable zealot, handles packaging and public relations; he is also the philosopher. At one point, midway through two days of oleic minutiae, he seized me by the arm. "It is simple," he said. "Making good oil is a series of small steps. If you mess up any one of them, your oil is bad, and nothing you do later can change that. The truth is in the nuances."

In Spain, where everyone else beats their trees with sticks, the Núñez de Prados insist on delicacy. Each season skilled crews from Seville finish picking table olives near home and hurry, some for the twenty-fifth year, to the family's four groves around Baena, in Córdoba province. The olives are picual and picuda, similar tart high-country Andalusian varieties which produce a flavorful oil that stays fresh for years.

Planning strategically, they pick only trees with olives that have just slipped from deep purple to black. The fruit is ripe and firm but not yet soft. Plenty of pressable olives lie freshly fallen under the trees, but they are left on the ground.

Six good harvesters can whack a tree clean in a few minutes. Hand-pickers take twenty minutes, if they work at lightning speed. Skilled fingers slide down a branch, detaching olives with a gentle milking motion. This protects the olives from bruises, which trigger acidity, and it protects the trees.

At the end of each morning and afternoon picking, the olives are trucked to the mill and pressed immediately. Most producers take pride in turning olives into oil within a few days, before fermentation begins. The obsessive brothers start to worry after two hours.

But the real difference is in the antique mill which the family acquired in 1795 from a local duke who had a monopoly in the region. It sits unobtrusively on a busy street in town, facing a park, behind a simple arched gateway wide enough for trucks. In 1989 Andrés and Paco refitted the old press and began to produce commercially in tiny amounts.

Because of the artisanal approach, their competitively priced oil probably costs ten times more to make than that produced by modern methods. It is better. That it is not ten times better does not seem to worry either brother. A few food writers caught on fast, and fancy markets in the United States and Europe clamored for all they could get. With no advertising, the company was making money, selling all the oil it could produce.

Paco started my tour with what he insists on calling a humble olive-man's breakfast in the decantation room, a shrine to olivedom. At one end of the long white tablecloth, a Serrano ham, complete with shank and trotter, rested on a wrought-iron rack. It was *pata negra*, the best there is, from a free-range mountain pig; the hoof is black rather than white. A clay bowl contained real capers: big green pods like pickled peppers. Those little round things we normally see, Paco explained, are buds of the caper flower. No comparison. Andrés brought in what he calls *huevos a la mala educación*, or ineptly done eggs. They were fried in—what else?—and cut into pieces. Among bouquets of fresh flowers and olive sprigs, there were cheeses, croissants, and pots of quince jelly.

As we sat down, Paco shoveled loaves of bread into an iron stove the duke must have used. "I love the ritual of this," he said, laughing slightly at himself. No one else was laughing. With the warm bread distributed,

Paco produced the *pièce de résistance*: two glass bottles of Núñez de Prado.

The first was last year's oil, mature and rounded, with a hint of ripe fruit behind the olive taste and a mild spicy finish. The second was for the real aficionados, a sharp, fresh oil, cold off the press. Rich in fresh polyphenols, it bordered on bitter, with that catch at the back of the palate the Italians call *pizzica*. We slopped both on everything: ham, eggs, bread, capers, cheese, laps, shoes. Much later, somewhat reluctantly, we got up for a look around.

Trucks dump the olives into beveled chutes in the lot outside, and they go underground to the crusher. At first, the brothers washed the olives. It seemed like a good idea at the time. "Later we saw no point in adding all that water to a natural process," Paco said. "The oil is better without it. We use no toxic pesticides and pick directly from the tree. Why wash?"

The olives are ground to mash by 300-ton granite cones, which are tuned by a specialist before each season. They are filed or chiseled for perfect balance and surface. This is an old Roman improvement on the *trapetum*, which crushed with two rounded stones, each the half of a massive granite sphere.

From the crusher, oily paste rides a conveyor to the Núñez de Prados' prized contraption, invented in Málaga last century by the Marquis of Acapulco. It is called a Thermofilter, though nothing approaching heat gets near any part of the process. It won't even warm the building when it is freezing outside.

The Thermofilter consists of two giant stainless steel rollers covered in a tight wire mesh, one atop the other, which lift and turn the paste at a snail's pace. Oil drips from the crushed olives by gravity, strained through tiny holes—50 per square millimeter—and runs out a trough at the bottom to a separate decantation system. This they call *flor de aceite*, the flower of oil. Not pressed, it just runs out by itself. Normally, pressing yields a kilo of oil for every five kilos of olives. One kilo of *flor* takes eleven kilos of olives.

The remaining mash is spread carefully onto *capachos*, the same double-layered *scourtins* used in France. Italians call them *fiscoli*, the name often used in California. Traditional mats are often woven from esparto grass; Tunisians make many of them. New ones are of synthetic fibers. But they are hardly different from the woven mats used at Ekron three thousand years ago.

At Núñez de Prado, a mill hand builds a tower of 120 mats loaded with mash, each placed like a doughnut on a central pillar and separated at intervals by metal disks for stability. The stack is trundled on a trolley to the press. The gasping hydraulic pump pushes up from the bottom, and oil runs down the tower into a spillway to the decantation tubs. It takes nearly an hour to compress by two meters. After each pressing, the mats are painstakingly cleaned so residue does not flavor the new oil.

Apart from folkloric splendor, the archaic press has decided scientific advantages. Modern systems crush the olives, pits and all, and mix in hot water to loosen the oil. If the water is not too hot, this is still considered a "cold press"; olive sellers and oil makers like to stretch reality. Then centrifugal drums spin the oil and water into separate pipes. This causes abrupt chemical change, Andrés insists, just as Max Doleatto did in Flayosquet. It aerates the oil, washes away volatile compounds, and separates the glucosides too quickly. Particles of humidity remain.

There is a price to superior oil, of course. By comparison, a nearby cooperative turns out an excellent oil that qualifies for the same DO label. Two workers there produce thirty times more oil than Andrés can make with a team of twelve.

The cooperative, named Germán Baena, is about as far as one can go to the other extreme. Rows of stainless-steel machinery glisten in the hard fluorescent light. Mill hands in matching coveralls keep watch in a control booth, studying the dancing red numbers and bouncing needles. With buttons or dials, they regulate water temperature, flow rate, and pressure. Finished oil is piped to storage vats. Spare parts and supplies are trundled around the large open room on noiseless rubber wheels.

Engineers pooh-pooh any molecular-level questions, stressing instead hygiene and efficiency. This battle of old versus new would follow me everywhere along the olive trail, growing steadily more complex. At this early stage, taste and tradition had me squarely on the side of the ancients. The Germán Baena process was no more exciting than canning corn.

As old as it is, the Núñez de Prado mill is so clean that you can, as the cliché goes, eat off the floor. Which is fortuitous, because that is exactly what you do. The new oil, black as overused Shell thirty weight, runs along open gutters. Slowly, it collects in the decantation room.

A dozen interconnected square tiled basins purify the oil, a process that dates back to Roman times. In the first basin, most of the vegetable water and impurities settle to the bottom, and the oil floating on top is piped to a second chamber. Waste water is drained off to be used as fertilizer for the trees. In the second basin, and others afterward, this separation by gravity continues for several days until only pure oil, cloudy and golden, is ready for bottling. *Flor de aceite*, clearer at the start, goes through four chambers before it is ready. Pressed oil takes eight.

To qualify as extra virgin, oil must contain less than 1 percent acidity. Fine Italian oils hover near .5 percent. Núñez de Prado ranges from .09 to .17 percent. This, in itself, means little, since such small variations in free oleic acid do not affect taste. But the low acidity suggests that flavor compounds have not been disturbed.

Acidity level is only part of it. Extra-virgin oil must pass a panel of tasters who grade its taste, aroma, and appearance. On an international scale of nine, an extra-virgin oil must score at least 6.5. For Baena producers, that is too low.

From the beginning, the brothers decided that filtering removed too much of the flavor. "It was a marketing problem," Paco said. "Many people who don't know much about oil like to see a clear liquid. They think that makes it better. But we felt our customers would prefer better oil. Besides, if we were the only ones with cloudy oil, that might set us apart."

From the final decantation basins oil is brought to another drafty chamber for bottling, Paco's domain. For tradition, he decided to put *flor de aceite* in the square-sided stubby half-liter glass bottle that families in Madrid used for bringing home bulk wine. Pressed oil is packed in five-liter tins. Later he also began using a few yellow porcelain bottles of the same shape.

Paco's assembly line looks more like a few dutiful sons helping put up the family preserves. Bottles on a table are filled, corked, and passed along. The label, a small booklet, is tied on by a string that runs under a plastic sleeve designed to prevent leakage. This is applied with an old paint-removing heat gun lashed to the arm of a flexible lamp. Next, someone melts red sealing wax onto the cork and imprints the brothers' seal. The Baena DO shield is glued to the bottle. Finally, the finished flask is numbered in India ink, logged onto the lined pages of a tattered book, and packed with five others in a cardboard box.

Each Núñez de Prado label, in whatever language, carries an invitation to visit the mill. They mean it. Good businessmen, they know that a look at their operation tends to turn casual consumers into faithful devotees. But it is the sort of Old World soft sell that went out with the Industrial Revolution.

Entertaining is done in the decantation room, where breakfast fades into the pre-luncheon apéritif and then high tea follows another round of apéritifs. The wine is Rioja and the sherry is the finest. Salami and cheese mysteriously materialize. The atmosphere is churchlike, and each morning someone puts fresh flowers on the altar, the red-tiled basin containing *flor de aceite*. At the slightest prompting, Andrés dips in a wineglass for ceremonial sampling.

But the real show is a ride to the trees. A two-lane blacktop threads its way into hills the color of red clay. Tucked back on the far rises, stone farmhouses crumble into ruin. Too many people have moved to town in order to earn a better living. Guardia Civil outposts have closed down for economy, and the families that remain are isolated and vulnerable in the remote countryside. Some now drive long distances to visit their olives, leaving ancestral homes to thieves and vandals.

Along the road, passing his neighbors' groves, Paco pointed to the bare earth between the rows of olives. "Herbicides," he said with a disapproving sniff. The brothers had found that over time a diminished use of chemicals meant fewer pests. Nature works that way, but only if everyone around has the same approach.

Most Baena planters have shifted to a refinement Andrés introduced in the 1970s. Before, they planted two shoots together so that twin trunks matured from the same root mass. This was to counter the olive's irksome trait of bearing heavily only in alternate years. But after Andrés changed his method, most now use only one shoot. The crown grows hardier and supports a stronger scaffold of main branches, and also produces more young olive-bearing wood. Careful pruning can keep production relatively stable.

The four Núñez de Prado estates total about one thousand acres. "We produce four hundred thousand kilos a year and cannot do more," Paco told me. "There is no more land." It is a family problem. His father had eleven brothers; his mother, ten. Altogether, about sixty cousins own huge tracts of land that currently do not produce for the company. "We want them to come in with us, but only if Andrés and I can control the process," he said. "They are all architects and professionals, and they don't see the value in spending so much money to do things the old way. They only think in terms of profit. Who knows? Maybe they are starting to understand."

He did not say it, but the future must cause some worry. Andrés is a bachelor. Paco's son is three. Two other brothers, like-minded but not in the business, have a few children between them. But the oldest son, just in his teens, has not yet made up his mind about olives.

We crossed the Guadajoz River, and the scenery changed. Lush grass grew among rows of handsome trees, all carefully pruned for uniformity and easy picking. Branches spread wide and dipped nearly to the ground. Expert hands had sawed away woody spurs to make room for new shoots. This was Andrés's territory.

"An olive tree has a lifespan like a man's," he explained. "At ten or twelve years it begins to produce. It matures, and its best years are between twenty and forty. After fifty it slows down, and after eighty it declines a lot. When a tree gets to be eighty, we pull it up to plant a new tree. We sell the wood for burning or making furniture."

Later, I mentioned to Paco my angst at allowing, through inattention, runaway vines to choke to death trees that were centuries old. He shrugged. "Nature is death and rebirth. That's the beauty of it."

We found the pickers deep in the grove. None slowed their pace at the bosses' approach. Hands moved in a blur, detaching olives and dropping them into baskets hung around the waist. Paco introduced me to Carmen, who first came to pick as a young woman. This year, her teenaged son had come along to help. Carmen was a two-handed picker; she could strip separate branches as fast as more workers could do a single one. She moved off too quickly for me to ask her last name.

"We believe in encouraging long-term relationships and loyalty," Paco said. "We pay more, but it is better in the end. Did you meet our accountant? His father was with our family for all his career. When he retired, his son replaced him. The packers, millers, all of them are proud of their oil."

Back at the mill, Andrés shot me a conspiratorial look. "I'll show you my secret project," he offered. When no one was looking, we slipped through a camouflaged opening in the back wall. Andrés had set up one of those fancy centrifugal mills, complete with digital readouts on a gleaming panel. It seems that a local operator had been buying up the olives they left on the ground, but the brothers did not like the deal. Instead, they would use the olives for a lower-quality oil, not sold under their name.

"It's going to make a lot of people mad around here," Andrés said with a little laugh. "This is the system they use to make their best oil, and we'll be using it to turn out junk."

We lunched at a tacky hotel, which had one of the finer kitchens in southern Spain. Paco had arranged for a sampling of a half dozen simple peasant dishes, all with Baena oil. The first was a sort of hearty gazpacho called *salmorejo*. In Andalusia, gazpacho is a hot-weather drink. *Salmorejo* is made similarly—a blend of the freshest vegetables on hand, with no water added—but with extra legumes to make a substantial soup. Other dishes were mixed with veal, pork sausage, cod, garbanzos, potatoes, eggs, and eggplant. The cook, the Spanish version of a Jewish mother, dished up each portion as if it were to last us until pruning time. She hovered nearby to make sure we enjoyed every bite.

I made a try at writing the recipe for *potaje del campo*, a muscular soup of garbanzos, chorizo sausage (Spanish, not Mexican), and vegetables. It began with sautéing garlic and onions and adding conical sweet peppers. But she lost me the fourth time she said, "You know . . . ," and flung out her arms. I should have realized—*potaje* means soup; *campo* means country. The recipe is the same around our neighborhood: Take what you have and cook it up for lunch. The only constant ingredient is olive oil.

After lunch, with some gentle probing, I tried to find out whether the brothers made much money with their thriving enterprise. Their production costs were enormous, and their prices were reasonable compared to what some Italian producers charged. Clearly, they weren't starving—but they were hardly flaunting wealth. Andrés clung faithfully to his perfectly maintained forty-year-old Mercedes-Benz with the original AM-only radio. Paco had his VW.

"I don't really think too much about the accounts," Paco said. "If I can live normally, I'm happy."

At 5:15 p.m., we roused ourselves from the table, long after I had planned to leave for Seville. Light would be fading, and I wanted to see the groves that were already old when the Romans arrived. "Just one more little *paseo*," Paco said.

He turned off the highway and wound up a narrow road into the hills of Baena. We were off to Zuheros, a whitewashed relic of the Moorish wars, one of those few Andalusian villages that have not changed in centuries. Ignoring a sign that said NO ENTRY, RESIDENTS ONLY, Paco inched up a paved donkey track past massive carved doors and balconies with wrought-iron grillwork. We peered through open gates at elaborate tiled gardens. At the edge of the village, we stopped to admire the crumbling stone towers of the *alcázar*, an old Muslim fort. Then he kept going up the hill.

Soon we rose above the olive line, where the last few trees struggled to survive among thick pines. Switchbacks angled upward until our VW was straining in second gear. Still, we kept going. Paco's running commentary melted into a reverential silence. We climbed on. Suddenly we stopped. I followed Paco down a footpath to a secluded overlook on a rocky outcrop.

Zuheros was a white patch far below. Baena was a pair of hilltop towers and a blur, barely discernible in the distance. From our perch, Paco showed me his view of olive heaven. We saw the entire valley, hill after hill nestled within a ring of low mountains along the horizons. Trees marched up and down in long, straight rows. The failing light flashed silver in some spots. In others it tinted the groves in shades of pink and peach and cobalt blue. And every square inch, except for a few ribbons of roadway and the odd clusters of buildings, was planted with olives.

———

Making good oil is one thing. But Jean-Pierre Vandelle, a friend of the Núñez de Prado brothers who owns El Olivo in Madrid, has made cooking with it into an art. A starworthy yet unpretentious French chef, he fell in love with sherry, Spanish oil, and Spain. His restaurant by a quiet plaza off the Castellana is a tribute to all three.

On each table dishes of olives and olive bread nestle among the crisp linen and gleaming silver. The *tapenade* is fresh and moist, with generous amounts of oil mashed in black olives, capers, anchovies, and garlic. One side wall is painted in olive sprigs, leaves, and fruits of favorite Spanish varieties: the pointed picual and the giant sevillano from Andalusia; the cornicabra from around Toledo; the tiny arbequina from Catalonia. A double-decker cart groans under the weight of fifty bottles of oil from all parts of Spain and a half dozen other countries.

"Not everything is cooked in olive oil, but almost," Vandelle said. "I have a friend in Bordeaux who has nearly perfected a way to make *mille-feuilles* with oil instead of butter. A very promising concept." He rubbed his hands together and imagined the flavor. "Tradition and respect for the proven ways are fundamental when it comes to olive oil," he added. "From that basis, you can experiment with what is new."

Different oils give each dish a subtle shift in flavor, Vandelle explained. One after the other, we plucked bottles from the cart for a guided oil tour of Spain. Vandelle likes the Baenas best, but there are a lot of fine oils. Sierra de Segura is a DO from the Andalusian province of Jaén. Like nearby Baena, it is sharp and yet mild enough to go with almost anything. But there are also the DOs from Catalonia, Siurana, and Las Garrigues (Borjas Blancas), made from different sorts of arbequinas.

The Siurana region is along the coast near Tarragona, where moist sea breezes produce spicy and light oils, with a hint of almonds. Gasull, from Reus, is rich in olivy aroma with a milder finish than Baena. Las Garrigues oils from the Pyrenees foothills of northern Catalonia are greener and more peppery. Lérida is a fine example. L'Estornell, also excellent, has a devoted following. Because the olive fly seldom ventures to high altitudes, its arbequinas can be organically grown.

Aragon produces small amounts of a mellow oil, prized by those who like it less green. Toledo brands are good, strongly olive-flavored, and certainly the fanciest; one comes in a painted round ceramic pot. I even liked

El Sublime, the top-of-the-line extra virgin with no fixed origin by Koipe, an Italian-controlled conglomerate. Careful blending had caught the full olive taste that Spaniards like without the greasy consistency that sometimes goes with it.

"People are starting to learn," Vandelle said. "Before, I would get very cultured clients asking me for our best refined oil." He gave a little shudder. Refined oil, extracted by steam and chemicals, is for making French fries.

I left Vandelle in charge of the meal. He opened with warm lobster chunks and salad greens, dressed sparingly with Baena oil. Then he brought his specialty, a *lotte* (monkfish) in tomato compote and a black-olive sauce, El Olivo. Later, he sent me the recipe.

For the compote, pare two pounds of tomatoes and remove the seeds and some of the juice. Mash the pulp. Brown two onions and three shallots in Spanish extra-virgin oil. Stew them slowly with the tomatoes, adding three cloves of garlic, sliced, and a sprig of thyme. Adjust the tomatoes' acidity with a pinch of sugar.

For the sauce, drop a diced yellow onion into hot oil and add twelve ounces of pitted black olives. Cook until liquid. Add two tablespoons of fish stock, some salt and ground black pepper, and reduce the sauce on a low flame. Crush the mixture and strain it until smooth. Finally, blend it in a mixer with a little good oil.

Grill two pounds of *lotte* with a drop of oil. Center the compote on the plate, with the fish on top. Surround the fish with a ribbon of sauce and decorate with parsley. That makes enough for four.

An excellent olive-oil meal in Madrid is easy to find. The simplest *tapas* bar, for instance, offers *gambas al ajillo*. These are small shrimps sizzled in oil with enough garlic to empty a large room; you mop up the remaining warm oil with bread. But the richest trove of recipes in Spain is in a soaring glass building, complete with guards who magnetically scan all visitors and X-ray their briefcases.

Olives have their own globe-girdling bureaucracy, based in Madrid: the International Olive Oil Council. Recipes, in fact, are a small part of it. The IOOC is an intergovernmental agency with twenty-three members:

the fifteen European Union states, along with Algeria, Cyprus, Egypt, Israel, Morocco, Tunisia, Turkey, and Yugoslavia. The United States is only an observer. Loosely linked to the United Nations Economic and Social Council, the IOOC has administered agreements on olives and oil since 1959.

"The IOOC aims for integrated development in the production, consumption, and marketing of olive products," its fact sheet declares. "Its activities encompass agriculture, technology, research, quality control, and promotion." Essentially, it keeps track of international flows, so that no one starts an olive-oil war. Twice a year, government ministers, specialists, and technicians fly off to some pleasant venue. Decisions, made by consensus, become mandatory.

Within its stuffy structure, the IOOC does some noble work. It keeps track of research, from growing trees in deserts to how people digest the last lipid in a drop of oil. Its library and data base are exhaustive. Its own publication index runs on for pages. Its magazine, *Olivae*, appears every other month, and regular bulletins detail production, trade, and consumption. Executive Director Fausto Luchetti, an Italian, is in constant motion.

Luchetti employs diplomacy and arm-twisting to get producing countries to work together. Competition is often fierce within the European Union, and also between the EU and other producers. Behind the scenes, national authorities complain that their rivals bend the rules, condone cheating, and disparage others to their own advantage. Subsidies and standards are often in dispute. Even where the IOOC has no enforcing powers, it is a backstage arbiter.

Beyond that, Luchetti tries to ensure a steady flow of high-quality, low-priced oil from everyone's mills to the pantries of the unenlightened. He is expected to prod those benighted masses beyond tasteless canned "ripe black" olives and pimiento-stuffed, lye-cured green olives.

Recipes are the province of Irfan Berkan, a jolly Turk behind an enormous desk who supplies a network of olive cheerleaders with press cuttings, new cooking ideas, and fresh approaches to promotion.

After a day of poking around the IOOC's inner reaches, I found myself sinking deep into detail. I was amazed to learn that countless microscopic hairs covers each pore on the bottom of a southern Algerian olive leaf. I thumbed through a report on high-performance liquid chromatography criteria for determining olive-oil grades. Then there was an arresting monograph on varietal polymorphism. Still, there were limits. I stopped at a paper entitled "Biometric and Protein-Enzymatic Characterisation of Some Olive Varieties," taking it home to alternate with Elmore Leonard.

But I had a question for Luchetti. Swept away by the IOOC's enthusiasm over the growing popularity of olive oil and its scientific frankness over the natural limits of production, I'd begun to worry. Was there ever likely to be a difficult year in which good oil might be in short supply? He laughed. "We should be so lucky," he replied.

During the first half of the 1990s, world oil production averaged 1,889,800 tons and consumption fell just below at 1,874,900. Carryover stocks provided a comfortable cushion. Table olives were in a more precarious balance. The average production was 970,400 against a consumption of 980,900, and sellers squeaked by because of stockpiled surpluses from an occasional bumper crop.

Less than a year after I spoke to Luchetti, however, 1996 turned out to be one of the worst ever. Oil production for 1995–96 fell far short of the average—about 1.56 million tons—with just enough carryover stocks to satisfy a projected 1996 consumption of 1.78 million tons. For table olives, the deficit was greater; even with stockpiles, demand would exceed supply.

As olive-oil prices climbed over 50 percent, marginal consumers dropped out in favor of seed oil and something else besides olives. Diehards bit the bullet and paid more. Over a few years, such a situation could get dramatic. But olives are cyclical—most trees produce well only every other year. Each region's crop depends upon weather that is seldom the same. Eventually the crops balance out, and lean years rarely come in pairs.

In the longer term, for oil at least, the IOOC declared itself optimistic. The rains had returned. Growers were planting more trees. A council study looking ahead concluded that for 2000–2001, production should be 2,070,100 tons and consumption 1,975,000 tons. The question was not so much whether there would be oil but, rather, where it would come from —and who would sell it. For the Spanish, this was a very touchy issue.

———

Juan Vicente Gomez Moya, dapper with slicked but thinning black hair, is a jovial ball of fire who runs a sixty-year-old organization called Asoliva. Unlike the IOOC, it is neither international nor neutral. A private trade group aided by the government, it is hellbent on selling Spanish oil. Each olive-producing country in the European Union has something similar (in Brussels, there is a Fedoliva). But Gomez has an especially tough job.

In a good year, Spain produces 600,000 tons of oil, up to half again as much as Italy, the runner-up. By comparison, France might turn out 2,000 tons. The Italians consistently run short, having to import oil for their domestic market as well as for the blended oils they sell abroad. According to the IOOC, Italy's average deficit is 140,000 tons a year. As this is based on dubious "official" tallies, the real figure may be higher. Spain might have a 200,000-ton surplus to export, but much of it goes out in bulk with no producer's label. Mostly, it goes to Italy.

And the imbalance could worsen. When the season ended in 1995, Spain's average production over a decade was 530,000 tons; its consumption 385,000 tons. By IOOC projections, in 2000–2001 Spain would produce 37 percent of the world's oil, compared to Italy's 19 percent and Greece's 17 percent. The Italian deficit of 222 million tons would equal Tunisia's entire production. Yet the Italians were still expected to tower over the American market.

Among consumers, Gomez explained, Spanish oil is woefully underrated. "We make the best oil anywhere, but people don't know it," he said with a well-practiced sigh. "Spain is the great unknown."

He had a simple explanation: The big market is the United States, and Americans have associated olive oil with Italy for generations. Italians are better at elegant labels and marketing. Half the sales are to restaurants and the food industry, which Gomez says are dominated by Italian-Americans. Supermarkets absorb much of the rest, and buyers stock what they know.

Gomez maintains that the basic patterns were established long ago, when Mafia muscle made offers that could not be refused. Now it is vastly more complex, he says, but Spain is barely in the game. When an oil boom hit the United States in the 1990s, Americans bought what was familiar. That was that.

Things were changing a bit, he said, and Asoliva labored to push them along. But a study by American consultants suggested an uphill struggle. Along with major Italian labels, supermarket chains sold their own brands at low prices. The strongest potential was in Florida, where Latin-American consumers liked the distinctive Spanish oil flavor. In the meantime, Spain focused on its traditional markets, where sales progressed at substantial rates: the Middle East, Asia, and Australia.

In all, about 470 Spanish labels are registered with the European Union, Gomez said. But many of the producers are small, selling only within their own region. Ten brands dominate 80 percent of the domestic market. The two giants, Carbonell (Elosua) and Koipe, are owned by Ferruzzi Finanziaria, a Milan-based conglomerate; their share alone is half the market. La Masia belongs to Unilever, which also owns Bertolli in Italy. Coosur is Spain's last state-owned enterprise. Pompeiian, an excellent blended extra-virgin oil sold widely in America, is from Seville. But most people think it is Italian. For independent-minded growers with pride in their identity, it is a tough business.

Talking to Gomez, I realized there was a tightly knit olive community scattered around the world. Everyone seemed to know everyone else: the scientists, the producers, the marketers, even the devoted eaters. Within Spanish circles, it was a family. When I mentioned Jean-Pierre Vandelle, he produced a gaily painted card sent out by El Olivo. It was a riddle, in formal Spanish: "Our faithful servant is green. His children are born white, but later they turn black. Who is he?"

We got out a map of Spain, and Gomez ran through the regions. He echoed some of Vandelle's preferences and added a few of his own. I was tempted to visit Reus in Catalonia. A group called Olís de Cataluña makes Oleastrum from arbequinas, hand-picked on 160,000 acres; it is owned by the Catalan state and cooperatives that represent 70,000 small growers. But my next move was obvious. With a few telephone calls I laid out a more thorough tour of Andalusia.

Before leaving, I browsed in Asoliva's little library. Paquita Vergara, Gomez's assistant, had put together a fine collection of oliviana. Her treasure was a tome by Miguel Herrero García, published in 1950, called *El Olivo a Través de las Letras Españoles*. It had nothing on the feuds and figures and the politics of men in pinstriped suits, but it explained why the Spanish made such a fuss. Clearly, money was part of it.

Olives have been the emblem of Spain since the first dispatches from Caesar's legions. When refugees in the Spanish Civil War cut down trees for firewood to survive, they mourned as they might deaths in the family. Republicans fought Francisco Franco's Nationalists over Belchite, in the heart of Aragon, and old oil mills and olive merchants' homes were destroyed. The ruins were left untouched, a monument to calamity. Today, olives grow everywhere in Spain, covering five million acres from the ancient port of Cádiz to the chilly slopes of Galicia. They have touched the Spanish soul and psyche for two thousand years.

It was all in Paquita's fat book, the poetry, the essays, the treatises. From Columella writing in Roman times, it went on through snatches of dialogue from Lope de Vega to García Lorca, Antonio Machado's "To Smile with Cheerful Grief of the Olive Tree," Don Quixote's asides, ancient sonnets, and modern odes.

A passage written in 1497, by some unknown hand in a forgotten dialect, was summed up in translation: "Olives are good, and healthy, and they stimulate the tongue. The young and old can eat them. Nothing is wasted, even the pits make charcoal. And their elixir awakens a sleeping sexual appetite."

In *La Dama de la oliva* (*The Woman of the Olive*), Tirso de Molina dramatizes an old legend: The Holy Virgin appears in an olive tree before a man condemned unjustly to the garrote. She tells him the tree will bear olives signifying poverty, obedience, chastity, and charity, the symbols of love and redemption held dear by the Order of Mercy. A monastery rises on that very spot. The man, nonetheless, is executed.

Lope de Vega laced his plays with olive references, harking back to *Betis olivifero*, Roman Andalusia, and reflecting bits of current folklore. In one scene, a character sings a verse still popular at harvest time in Soria province:

> *Ay, fortune,*
> *Pick me that olive!*
> *Satiny olive,*
> *Outside green and tender,*

And ripe on the inside,
Fruit hard and welcome
Ay, fortune,
Pick me that olive!

Popular *zarzuelas*, those danceable musical comedies so beloved by the Spanish, celebrate the noble little fruit. One of them goes:

So deeply you carved your name
That the olive tree was lost:
If you should ever forget me,
What a shame for that little tree.

Federico García Lorca's work was rich with olives. In his poetry, Spain was symbolized by its olive groves, and his tree had life and personality. There was, for instance, "Landscape" from *Poema del Cante Jondo*, which I later found translated by my friend Willis Barnstone:

The field
of olive trees
opens and shuts
like a fan.
Over the olive groves
lies a sunken sky
and a dark rainfall
of cold stars.
Penumbra and reeds tremble
on the riverbank.
Gray wind ripples.
The olive trees
are loaded
with shouting.
A band
of captured birds
moving their broadest
tails in shadow.

After a while I stopped and wondered if anyone would mind my lighting a cigar. Silly question. A massive ashtray on the low table held the stub of the last visitor's Cohiba Lancero. The room was perfect lived-in Old World ornate. Rococo moldings framed high vaulted ceilings. Antique armoires stood against dark-wood paneling. In the doors, etched glass was set in leaded panes. A painting, a stylized montage of olive-oil labels, dominated one wall. A tall sculpture in ancient olive wood filled a corner by a long leather sofa with huge puffy cushions. Men in perfectly cut suits and real shoes strode the creaking parquet. Paquita bustled about, a classy woman of a certain age, trailing cigarette smoke and ribald remarks.

There was something about the seedy elegance of that room and its setting. In a tiny park near the building, a handsome old tree defied Madrid's traffic fumes. Maybe I was simply enjoying the company of people who argued passionately about centuries-old literature, knew their wine, considered a two-hour lunch to be fast food. Perhaps it was all that reverence for good oil and its range of possibilities to enrich a life. Spain was olive country down to its deepest roots, and this quiet room felt close to the heart of it.

Shapes and Sizes

Ropa Vieja

At Casa Juanito in the Andalusian town of Baeza, Juanito and Luisa offer an appetizer they call *ropa vieja*. The name, meaning "old clothes," is no description of the taste. A sort of Spanish *hummus*, its flavor depends partly on the sharp, fruity oil of the high country in southern Spain.

1 pound dry garbanzos
1 small chicken leg, skinned (optional)
1 thin slice of veal or lamb (optional)
1 small piece of ham bone (optional)

2 onions, very finely sliced
1 cup extra-virgin olive oil
4 ripe tomatoes, peeled and diced
Salt

Soak the garbanzos overnight. Drain and cook them with the chicken leg, veal or lamb, and the ham bone for 20–30 minutes in a pressure cooker or 1–1½ hours in a large pot. (The meat adds flavor to the garbanzos, but may be omitted.) Mash the garbanzos and set aside. Fry the onions in oil until tender and add the tomatoes, keeping the pan on low heat until the tomatoes blend well with the onions. Stir in the mashed garbanzos, mixing well to a homogenous texture. Add more oil if needed. Salt to taste.

"Andalusians of Jaén
Grand oilmen,
Tell me from your soul, who
Who brought the olives?

"Nothing brought them,
Neither money, nor the Lord,
But the silent land,
And struggle and sweat."
—Miguel Hernández

5

Olives of Wrath

It is dark, cold, and late at the center of the olive world, a town called Martos in the Andalusian province of Jaén, and the Sunday whackers wait by their loads. They stand for hours as the line inches forward, their stoic Spanish faces in endurance mode. Most began picking at first light, and it is approaching midnight. One by one, vehicles pull up to a hole in the ground—Land Rovers, tractors, toy-size station wagons, dump trucks—and spill out olives that will make some of the most popular Italian oil.

After their loads are weighed, the men collect their chits and go home to dinner. That is the last they'll hear of the olives they worked all year to grow, except for two checks which will turn up in the mail. The co-operative will give them their share of sales, based on the market price. And the European common market will pay its growers' subsidy.

Much of the oil will be sold to Italian companies, which also buy oil in Greece, Tunisia, and Turkey, to be blended and labeled *Product of Italy*.

And much will go to major Spanish packers which are controlled by Italian conglomerates.

Among themselves, Jaén growers grumble mightily at the system. Their province alone produces about a tenth of the world's olive oil. Around Martos, where the groves are most heavily concentrated, their honor is involved. While some of their oil is packed and sold under its own name, most is shipped off anonymously in bulk. Under European Union rules, anyone can buy oil within the common market and move it across internal borders with no restrictions.

Here was the reality of what Juan Vicente Gomez Moya had told me in Madrid. Italians dominate the crucial American market. But since they eat almost all their own good oil, the Jaén growers know, they have little to sell.

This Italian run on Spanish oil is hardly new. Elderly Spaniards re-member, not fondly, the World War II occupation, when Italian packers charged a premium at home for oil labeled *From our colonies in Spain.* Last century, archaeologists found that Monte Testaccio on the Tiber was not a Roman hill but rather a dump for at least forty million broken oil am-phoras from Spain. The seals were dated from A.D. 140 to 265.

What really hurts now, the growers say, is that Italians get not only the glory but also the European export subsidies. With artful shuffling or outright corruption, many declare that the Italian oil they buy outside the country comes from nonexistent trees in Italy. And while European Union auditors routinely exposed outrageous olive scams, officials in Brussels have few means to stop them.

A spirited weekly in Madrid, *Olea*, editorializes regularly on what it calls Italian mislabeling and other fraud. Twenty percent of the "pure olive oil" packaged in Italy is cut with colza oil, asserts editor Vidal Mate, who is also the agriculture writer for the daily *El País*. Italians, in turn, accuse the Spanish of doing their own cutting.

The rivalry is bitter. When Ferruzzi began what would be a two-year campaign to absorb Elosua in the early 1990s, the Popular Party tried with-out success to start a parliamentary investigation. "Elosua had received a huge state subsidy to keep it in Spanish hands, and then they turn around and talk about selling to the Italians," party spokesman Miguel Ramírez

Gonzales told me. "This is a matter of national pride," he said. Ramírez said bribes were involved, but nothing was proved.

When the government in Madrid finally gave up its objections to the controversial takeover of Koipe by Ferruzzi, the opposition cried foul. Parliament members said the Italians had bribed Spanish politicians. When I telephoned Davide Pozzi, the Ferruzzi spokesman in Milan, he chuckled. "It wouldn't surprise me," he said. The whole Ferruzzi directorate collapsed in 1993, mainly because of bribery scandals.

The Spain-versus-Italy issue is a rueful joke in Madrid. According to one set of IOOC figures, Spain grows 23 percent of the world's olives, and Italy 22 percent. When I mentioned the numbers to Gomez at Asoliva, he coughed with laughter. "Yeah?" he replied. "Go to Italy and count the trees."

This is also a sore subject in Italy. Italian producers who bottle excellent oil under their own names suffer when someone refers to "Italian" oil in disparaging terms. Reputable blenders insist that, national pride aside, a mix of different sorts of quality oils yields the best product. They excoriate countrymen who buy inferior oil, from Spain or elsewhere, and sell it as extra virgin.

But in Jaén, hardly anyone talks about such things for the record. Growers told me that a local trade group had made a sampling in the United States. It took ten blended Italian oils off market shelves and analyzed them. Only one lived up to the claim of its label. This was purported proof that Italians could not be trusted. No objective observer oversaw the sampling, however. The results were not published for fear the Italians might accuse them of cheating.

In Jaén, where olives are big business—and the only business—pride gives way to pragmatism. Oil is a cash crop and not much sentiment can be wasted on it. Some people make fortunes. A fair number do well. A lot more scrape by at the edges.

Francisco Molina, the Jaén producers' spokesman, took only a guarded swing at the Italians. "Spain is the world's main producer," he said. "We have the highest level of technology, the most investment, and the greatest rate of reinvestment, but we lack a commercial sense. The Italians are able to manipulate things so that our traditional markets are awash in 'Italian' oil. There is no national oil, no such thing as an Italian type or a Spanish type. It's from a plant, with many varieties, all different, with distinct climates. Consumers can't tell which is which."

Although the same European Union regulations apply equally to all fifteen member states, Molina claims that Spain enforces them rigidly while Italian authorities do not. "We face unfair, illicit competition," he said. "I can't compete if you give me a sword and the other guy a rifle."

Molina's organization in Jaén is part of a network which helps administer EU subsidies of dizzying complexity. Each year Brussels fixes a minimum price for olive oil, compensating farmers in case of a shortfall. It determines a subsidy in the range of a dollar a liter of oil (or a larger amount per tree, if production is small). It pays a per-kilo subsidy to large exporters. And it helps olive growers in other ways.

Most producing areas in Spain, Italy, Greece, and Portugal have a local group, which belongs to a larger union. These unions handle much of the paperwork for Brussels. But the EU's aim is a minimum fair deal for everyone, across the board. The producer unions seek the maximum. Subsidies aside, they look for competing customers who will bid up the prices for their local oils.

Molina's Jaén Olive Producers' Association supports an offshoot organization run by a young northerner named Jesús Cuervas García. Its purpose is to encourage Jaén growers to package oil under their own name and sell directly to overseas markets. Cuervas, an outsider, was the exception in Jaén. He spoke with great patience, opened files, and later sent me thick background reports. Unlike others, he was blunt about Italy.

"The Italians are charming," he told me. "I'm enraptured by their culture. Now, here is a people who love their olives. But they are the biggest liars in the world. They know that you know they are lying; it's their world, their life. This allows for absolutely incredible, unbelievable fraud across Italy."

The fault, Cuervas said, is not solely with the Italians. Local growers should band together to enforce a higher selling price. And they should put their own product on the market. "The problem is that no one knows he has a problem because they're all making too much money," he said. "Ten years ago they came to the mill with a donkey, and now they have two cars in the garage. The Jaén grower is making what he never dreamed of earning from olives. So what if he produces and does not know how to sell?"

Cuervas knows that growers cannot count on easy bulk sales. We spoke as producers delivered their 1994–95 crop. After four years of drought, the yield was still good. But almost no rain fell over the next year, and many trees came close to dying. Production for 1995–96 was a fifth to a quarter of its average. Italian buyers scrambled to find new suppliers in Greece and Tunisia. Then a winter downpour—the most abundant in 130 years—cheered Jaén oilmen. And yet if the following year's production was too good, prices might drop substantially.

In the end, Cuervas said, there is no other way. Unless the cooperatives can establish their own direct sales, with their own labels, they are at the mercy of nature—and fickle bulk buyers. But he is hardly optimistic.

"I, who fight every day to get these people together, don't see it happening," he concluded. "And this is what allows the Italians to control Spanish oil."

Cuervas's analysis seemed to explain why so few Jaén growers wanted to point fingers. Their reticence was described to me as politeness. It was not nice to criticize a European partner. Another guess might be that, with no structure in place to compete in the open market, they were happy enough with a fat check when the going was good. When production was down, they did not want steady buyers to forget them. In such circumstances it is not wise to irritate gilt-edged customers, not even if they are Italians.

The medieval town of Ubeda lies northeast of Jaén and west of the Sierra de Segura. Halfway to Ubeda, past hills neatly terraced with rows of olive trees that disappear over the horizon into La Mancha, I stopped to chat with olive whackers. A black plastic mesh tarp, a *manta*, was spread beneath the tree to catch what fell. Within three minutes of attacking a fresh tree, the whackers were ankle deep in delicate green foliage and olives that mashed under their feet. Then they moved on, and two women hauled the *manta* after them. Two more scrounged stray olives from the ground.

These were freelance pickers. During the year they worked in fields or restaurants, or nowhere at all. For two winter months they picked six and a half hours each day, with a short break for chorizo and warming wine from a straw basket. The men earned 4,055 pesetas daily, or $30; the women made 3,981 pesetas, fifty cents less. Growers pay by time, rather than weight, so that whackers do not ruin trees by working too fast. They get their money's worth.

Cristobal García, a young man in a blond Lenin goatee, answered my questions in polite monosyllables. He was not about to be distracted. Apart from picking, García scraped by with odd jobs. "In Jaén," he said, "there is only olives." I asked who owned the trees, and he shrugged. Some guy in La Mancha, he thought. For that, I should ask the foreman.

That was Fernando Valverde, at the wheel of a big yellow tractor. A full-time employee, he helped take care of the estate's twelve thousand trees, mostly hojiblancas. About half were irrigated by black drip tubes.

"Who owns this *finca?*" I asked.

Valverde looked genuinely puzzled.

"Um," he said, after some pondering. "He lives in Jaén, I think, but I can't remember his name."

Ubeda, high on a hill, flourished with olive money soon after the Moors were expelled from southern Spain. Noble families built splendid mansions and soon fought so fiercely among themselves that in 1503 King Ferdinand and Queen Isabella tore down the town's ramparts to instill the fear of God.

On a back street of Ubeda I found an old man and his son sewing up a *manta* like fishermen mending their nets. He had six thousand trees. When I asked if he could make a comfortable living from them, a huge grin spread across his face. "Oh yes," he said.

He hired outside help only for the harvest. "You need good people for that," he said. "If they don't know how to use the poles, they can ruin your trees." He was considering an innovation, a tractor-powered vibrator that shook olives off the tree, saving manpower. He gave his name as José. Asked the rest, he shrugged and returned to his *manta*.

I thought of José a year later when Jaén olivemen made *The New York Times*. The drought's fifth year, the hardest of all, had been difficult for everyone. Some growers brought in less than 10 percent of their average crop. Smallholders who depended on their olives would be adding water to the soup all winter long. Pickers paid by the hour would be in desperate straits. But, of course, the rains came again.

On a winding cobblestone street, I found Father Antonio, a twenty-nine-year-old priest from Extremadura, in the back reaches of San Miguel Church. Despite the olive wealth, he said, Ubeda and the surrounding regions were calamitously short of work, and he worried that many large tracts of land had remained undivided. "Future prospects are directly related to olives. That's all there is. Big families have the riches but don't know what to do with them. I see no creativity to use what they have to make something better. The big companies come from Catalonia, Madrid, Italy. They take away the olives, leave some money, and that's it."

As far as Father Antonio knew, it has always been that way. "Some of these fortunes go back four hundred years," he said. Over centuries, scions married elsewhere and moved away. Original clans, spread across Spain, are hard to track on paper. But, the young priest concluded, they make up a conservative ruling elite. "There are four or five big families, and nothing more."

This was Steinbeck, European neo-feudal style. Working flat out for three months, Sundays included, a picker earns $2,700 before deductions and living expenses. He has no reserve for a bad year, when there is nothing to pick. A landowner might have fifty thousand trees, if not four or five times more. In a decent year, that is perhaps 400,000 kilos of oil. Labor, pesticides, and milling add up. But a yearly profit approaching a million dollars is not bad for an absentee landlord.

Along with the big holdings, huge tracts of olive country are owned by small-timers. A clerk at Jaén's City Hall has two hundred trees he tends on the weekend. Dentists and doormen own trees as a hobby, a passion, or, more commonly, a way to get through to the end of the year. These growers each belong to one of more than a thousand cooperatives, which press and sell their oil. The Sociedad Cooperativa Andaluza del Campo Domingo Solís is one of the biggest.

Out front, I found dozens of farmers standing around waiting to check in their olives. Spaniards, by and large, like to talk—a reporter's dream. I sidled up to a likely pair leaning against a trailer. Both were pleasantly communicative. When I asked about olives, the topic shifted gently to the weather. After a while, they resumed their conversation as if I had gone.

Down the line, the same thing happened. Philosophical questions got answered, philosophically. Direct ones—how many trees do you own?—hung heavily in the cold air. My Spanish was Mexican-Argentine; no *z*'s softened in my teeth. Perhaps they took me for an English spy from the European Union, collecting dirt for the next round of subsidy negotiations. Possibly I was an Italian, up to nonspecific no good. For whatever reason, these Jaén weekend olivemen were taking no chances.

Baranca Teba was shivering by himself near his clapped-out Land Rover. He'd been waiting an hour to dump a pitiful few sacks of olives, and the line was still long; he looked desperate enough to talk to anyone. Teba worked year round in other people's fields. On his days off, he looked after his own two hundred trees. He picks his crop over five or six Sundays and, after five or six waits in line, sends his olives off on an anonymous journey. Not tasting his own oil causes Teba no existential problems.

"If the year is good, I might make 500,000 pesetas, maybe 600,000," he concluded, before turning away to contemplate the empty sky. That is between $3,700 and $4,500. "It's something."

Inside the mill, Francisco Gutiérrez López bustled about in blue coveralls with a screwdriver sticking out of his pocket. Thirty-one, with an easy laugh, he was thrilled with his work. Francisco originally trained as an auto mechanic. He married the daughter of the mill's *maestro* and discovered he loved the business. He started as an apprentice and soon advanced to foreman. When his father-in-law retired, he was made the boss. Francisco's title was *maestro*, but it might have also have been commodore. The mill looked like the engine room of the QE 2.

At full blast the mill churns out 200,000 kilos of oil in twenty-four hours. In a single weekend it can produce as much oil as the Núñez de Prados make in a season. Normally, from December to March, it digests ten million olives. The machines are Pieralisis, those ubiquitous Italian centrifugal grinders I saw in the West Bank. A licensed factory makes them in Zaragoza.

Francisco walked me through the process. It is similar to Beit Jallah's except five banks of huge machines mill at once, side by side. As was usual for modern milling systems, everything was stainless steel, inside and out. This prevents oil from picking up a metallic taste. Also, surfaces can be polished to a blinding gleam. But grimy metal panels on the floor covered the greasy pulleys, gears, belts, and cables that made the system run.

Weighed and washed, the olives are mixed with hot water and crushed to mash in the giant horizontal drums. I watched the temperature gauge climb toward 120 degrees; this was pushing it hard. Chemists say oil is altered above 85 degrees. Black goo is pumped into vertical spinners to decant at high speed. Finished oil flows into collection vats and settles under a murky froth.

In a final step, the oil is pumped to steel cylinders, where it is filtered for purity and clarity of color. Francisco lets the oil sit for weeks in stainless-steel storage tanks to reduce bitterness. After milling, a fair amount of oleuropein stays in the oil. It can be reduced by washing, he said, but that costs too much. For flavor's sake, that is just as well.

In the laboratory, a young woman named María Carmen analyzed the oil content of every batch to make sure cooperative members were paid accurately. We looked at the growing mountain of gray-brown pomace, *orujo* in Spanish, that is sold to refiners for low-quality oil. Finally, Francisco took me into the boiler room, where the last bit of olive waste is trucked back again from the refinery and burned to heat water for the process. The perfect circle.

At around .2 percent, the final acidity is not much more than the best of Baena. With all the hot water added to the process, however, subtle aromas and flavors are missing. Still, Francisco savors his oil. He held up a clear glass of it to admire the color. "A lot of affection goes into this," he said. That morning, as usual, it was his breakfast: a generous drizzle on a piece of bread. At lunch, he would eat it the same way, but with salted cod, some tuna, or a fresh tomato.

I asked Francisco what he thought of his oil being hauled off to Italy. He was not afflicted by reticence. "I get furious," he said. "We are the best producers in the world and have to settle for the second rank."

He said 99 percent of Jaén oil goes to the Italians, allowing himself a little olive license. "You cannot make a lot of good oil in Italy," he said.

"It is a long, skinny country, and too much of it is close to the coast, vulnerable to olive fly. They have nothing like we do, a big region, high up and far from the water. Italy can never produce oil like Jaén."

Francisco gave a helpless shrug. "What can we do?" he concluded. "The Italians are so well situated in the market, you couldn't move them with an earthquake."

As a city, Jaén is not much to see. But Andalusia is mostly magnificent: rich in color, noble customs, a tradition of hospitality and tolerance. It is wildly diverse. Seville, urbane and well watered, is among the world's most thrilling cities. Almería, cut off by miles of rocky desert, has the sleepy air of a backwater Mexican fishing port. Granada and Córdoba are dramatic, rooted deeply in history. Flower-splashed villages climb impossible hills; their church belfries and Arab watchtowers are visible for miles. Isolated farms hide low in the terrain. And throughout Andalusia, the unifying theme is olives.

Slanting light through the trees plays shadow games on Andalusian roads. Young couples escape the parental eye deep in the old groves. Oleaginous references flavor the language. Breakfast eggs are cooked in olive oil. More often, breakfast is only oil and bread. Even in the fancy dark-paneled cafés of Seville, where people belly up to a brass bar for morning coffee, cruets hold sharp, fresh oil to anoint toast.

I traveled through Andalusia with Jeannette Hermann, who shares Wild Olives with me. A professional traveler who had never seen Spain, she ooh-ed and ah-ed as we rounded bends and saw yet another vista of silvery green. From Jaén we had a specific destination. "Do not miss Casa Juanito in Baeza," Vandelle had said in Madrid. "You will like the food." From him, this carried the weight of a papal decree.

Baeza is a little town near Ubeda, a labyrinth of medieval lanes still used by donkeys laden with olive branches cut as goat fodder. On the outskirts we saw JUANITO painted on a small hotel-restaurant. It was not much on the outside. Inside, a vast expanse of white tablecloths was overseen by a band of seasoned waiters. Well into the lunch hour, at 3 p.m., not a chair was empty.

Juanito turned out to be a short, bald dynamo with a satisfied look that can develop into a pleasant leer. Luisa, his wife, seemed a perfect match, although her easy smile was demure. As an *amuse-guele*, a light appetizer, they offered something called *ropa vieja*, which means "old clothes." It was garbanzos mashed in oil with lightly sautéed onions, diced tomato, and a tiny bit of salt. As the name was meant to suggest, it was a simple, familiar Andalusian taste, comforting as a well-worn shirt.

With some lovely Rioja red wine, Juanito talked about olives. "I can't tell you about all that antioxidant stuff," he said. "When we get into that at seminars, I tune out. We each have our expertise: Cobbler, stick to your last. What I know is taste." He explained that Baeza's rocky, well-drained soil and long, dry summers produced small olives that were rich in tasty oil. It was different near Seville, where more moisture meant better table olives. Each region had its own characteristics.

Juanito thought it was best for cooks to stay with one oil they know and like. His came from a small Baeza cooperative that sells only unlabeled five-liter cans. He laughed hard at how clever packagers can make so much money from oil that is not particularly superior.

"The richer the people," he said, "the dumber they are. This virgin and extra virgin is to confuse. A woman is a virgin or she is not. How can she be extra virgin? What matters is taste."

He said Spain's entry into the European Union had improved its oil but not necessarily growers' earnings. "We are very bad businessmen," he said. "Eight, nine years ago, the state bought all the oil. If it was bad, they paid very cheap. Now competition makes better oil. But last year the Italians took it all. It went to the Mafia and the multinationals."

Juanito, robust at sixty-three, claims to consume a quarter liter a day of oil. In the morning, he fries an egg in it. He pours it on his toast with *café con leche*. For dinner, he takes a small round loaf, digs out a hole in it, pours in oil, and sprinkles on a little salt.

"Look at us, strong," he said, giving Luisa a rap on the shoulders. He ticked off their roster of healthy children and grandchildren. "Do you have children?" he asked. Like Farid, he took the negative reply sadly. He turned his attention to Jeannette. "Use more olive oil and you'll have those children," he said. It has, he posited, distinct properties as an aphrodisiac.

Lunch was stretching toward three hours. The 1985 Yllera had given way to homemade fruit brandy. Juanito rhapsodized on the joy of cooking.

"Never measure," he said. "You must never measure. Splash. Pinch. Use your eyes, your instinct. Never spare the oil." What did he think about butter people? "*Tontos*," he replied. "Fools."

When we got up to leave, finally, I asked Juanito, "Are you sure you never cook anything in butter?"

"Feh," he spat. "Poison."

Italian Frantoio

Pollo al Mattone (or al Diavolo)

From her home in Tuscany, Nancy Harmon Jenkins roams Italy for recipes. Among her favorites is Tuscan grilled chicken, flattened by a brick (*mattone*) and hot as the devil (*diavolo*).

1½ cups extra-virgin olive oil
1 teaspoon salt
1 tablespoon freshly ground black pepper
1 tablespoon hot red pepper flakes, or to taste
1 tablespoon chopped fresh rosemary leaves

1 tablespoon finely chopped flat-leaf parsley
1 tablespoon fresh thyme, or 1 teaspoon dried, crumbled
3 small chickens split in half, or larger ones in 6 or 8 pieces
Juice of 1 lemon and lemon wedges for garnish

Mix the olive oil with the salt, peppers, rosemary, parsley, and thyme and set aside, covered, for at least 2 hours. Pare the excess fat from the chickens; flatten by laying them skin side down on a board and pounding them smartly with the side of a cleaver. Place the chickens on a large platter and cover them with the flavored oil. Marinate for at least an hour, turning occasionally. Set the chicken on a hot grill, skin side down, a good 8 inches from the fire. Add the lemon juice to the oil left in the platter and baste the chickens as they cook. Grill 15 minutes on each side, turning each piece once. The juices should run clear yellow when thoroughly cooked.

Serves 6–8.

Adapted from *Mediterranean Flavours*

*"And with the sprig of a fruited olive
man is purified in extreme health."*
—Virgil, Aeneid

$$\overline{6}$$

Top-of-the-Line Tuscan

In a two-room office in Florence, Piero Tesi held up an oversized perfume bottle. It was, he claimed, the vessel in which an elite group of central Tuscan olive growers would soar to glory. Vintners in Burgundy and Bordeaux had long since settled into their niche as the world's finest winemakers. But no one had yet been universally recognized as makers of a surpassing *grand cru* olive oil to equal that cachet. This would be Laudemio.

Tesi is director of Laudemio, a consortium of independent estates. The name is an old term for the tithe serfs paid to their lord, the best of their crop. Early success, if limited, demonstrated an Italian mastery of the two keys to selling quality oil. They made a good product, clear and leafy green, with the sneaky sharp afterbite—*pizzica*—that certain connoisseurs love. And they carried marketing flimflam beyond imagination.

Estates within a carefully delineated area are entitled to approach the Laudemio altar. Their olives must be grown only on the Colli Della

Toscana Centrale, the hills between Florence and Siena defined by the Italian Ministry of Agriculture in 1984. To be blessed, a producer must comply with guidelines set forth in an arty brochure that comes with every bottle. They have to be picked by hand, before December 15, and cold-pressed with a "reduced amount of time" twixt tree and mill. Harvesting and processing are monitored.

If the oil passes its test for acidity and organoleptic properties, it is packed in Laudemio bottles, with the estate's own name discreetly added lower on the label. Each half-liter bottle sells in the United States for nearly thirty dollars. By comparison, a five-liter can of Núñez de Prado costs seventy.

The strategy is to seize the high road. The thirty-five Laudemio producers call themselves Gli Olivanti, which is an elegant Italian way of saying Olives "Я" Us. Except that their only product is olive oil. Gli Olivanti grow frantoio, moraiolo, and leccino olives, which produce their distinctive flavor. But so do others who are not in the group. Neither guidelines nor prices can ensure that theirs is better than their neighbors'.

Like wine, olive oil varies by the year; depending on the rains, it can be rich or watery. Unlike wine, estate-bottled oil cannot be made better with blending and biochemical fanciwork. It is, essentially, only juice pressed from a fruit.

The brochure says a fixed harvest date is to avoid using "olives which have gone beyond the optimal stage of ripeness." But olives follow the weather, not the calendar. They ripen at different times each year. As they are hand-picked, overripe olives can be left aside. Tuscans pick in November, sometimes even in October, because they like their oil leafy green and spicy. In any case, their hills are colder than most olive-growing areas. By mid-December, most trees are clean.

Tesi, an impassioned but honest man, acknowledges these elements of nature. "Wine is to religion," he said, with a Florentine flourish, "as olive oil is to philosophy." The latter leaves out mysticism and a certain act of faith. Nonetheless, he fondled the faceted flask as if it were the Golden Calf, and he proselytized with conviction.

"There are microclimates in each region of production, soil characteristics, which is the reason for *les appellations d'origine controlée* in wines," he said. "The process must be constantly perfected. We believe in respect for the traditions, not the bad habits. This way, we are able to produce a voluptuous oil, to be identified with the highest levels of grand gastronomy."

Tesi said the oil is spun centrifugally because that homogenizes the oil better. A lot of chemists would dispute that point, and some traditional oil makers would fight a duel over it. But I did not argue. For large volumes, anything but a centrifuge is slow and expensive. Laudemio is also strenuously filtered to be perfectly translucent, with no trace particles.

Gli Olivanti share the costs of a common laboratory, which seeks ways to improve oil quality, as well as specialists who try to grow more productive trees. And, of course, impressive hype.

Their packaging works. Before visiting Tesi, I stopped in a well-stocked wine and oil shop down the street. My eye traveled straight to the classy cream-white box with a Roman athlete broad jumping across the front. I opened it and removed a crystal-clear flask. Now who, I wondered, would pack oil in a bottle that practically screamed for the sunlight to ruin its contents? And who had the nerve to charge that much for half a liter? I'd bite. I peeled off 33,600 lire, close to thirty dollars, and walked out with it.

Months later, I noticed an identical box at Tower Market in a not particularly affluent neighborhood of San Francisco. The price was twenty-six dollars. I found the manager, an affable man who loves talking to customers, and asked him if many people paid that kind of money for Laudemio. "Yes, a lot," he said, eyes narrowing, as in "You wanna make something of it?" He added, "It is a very good oil."

As it turned out, my Laudemio, and Tower Market's, was from the Marchesi de Frescobaldi, whose wines have been celebrated around Florence since Giotto and Botticelli were painting olive trees. Family fortunes, dwindling somewhat, were improved when Vittorio Frescobaldi married Bona Marchi, a woman of legendary beauty from an industrialist clan that raked in legendary profits. La Marchesa Bona turned Frescobaldi olive oil into big business.

At her office off the Piazza Frescobaldi, the marchesa, a blonde of commanding presence, exuded elegance. Gold dripped from her wrists and, in long chains, onto her tailored black suit. She patiently let me exercise my pathetic Italian until I shifted to a more comfortable French. Then, politely but pointedly, she switched to perfect English, with a posh British accent.

Like most oil converts, she was a gusher on the subject. At the first Oleum olive fair in Florence, in 1994, she had put together a rich display of her *objets*: antique cruets, vessels, carvings, silver spoons for serving olives.

"Think of what olive oil has brought to our culture and our nourishment," she said. "Everybody in our wealthy world is trying to lose weight. Laudemio is excellent when on a diet. You can use it alone on raw vegetables or boiled vegetables. You don't even need to add salt, because the taste is so strong."

With so many oils on the market now, she said, people are confused. "This product is a guarantee for the consumer. It is the same idea as with the French *grand cru* wines. You know what you are getting."

But do you really know? I asked. Well, she allowed with a diffident chuckle, at least you can be sure the producers did the best they could with what they had. "We don't always produce the same quality," she said. "I had a bottle of Laudemio the other day that was positively vile." And, she added, the best oil can turn bad quickly in the sunlight.

Laudemio producers, like all oil makers, faced a packaging dilemma. If they used a tinted bottle, consumers could not see the clear, fresh color. They decided to add the cardboard box so the oil was visible but also protected. But many people simply throw the box away.

Just as Tesi had, the marchesa emphasized the difference between grapes and olives. "Wine is greatly influenced by the hand of man," she said. "With oil, it is different. The olives carry their own potential, and all you can do is stay out of the way."

Months later, at the 1995 Oleum fair, I returned to the Piazza Frescobaldi for a cocktail party in the private apartments. A score of Laudemio oils were arrayed on a table next to baskets of bread cubes. I sampled one after the other, surprised at the differences. That year's Frescobaldi was green and grassy. I preferred Marchesi Antinori and Baggiolino, both farther over the subtle line between peppery and bitter. The Santedame was sweeter and fruitier. And all of them were as assertive as sword swipes.

Padding across the Old World carpets, taking in the paintings and paneling and high walls of aging leather volumes, I saw the larger perspective. These families were pressing olives when Medici and Machiavelli were hatching plots. The next day, at the Frescobaldi estate at Nipozzano, Bona's daughter Diana led me under carved stone archways, through a tumbling-down medieval village, among trees with roots in the Renaissance. Tradition oozed from every crevice. In a dank wine cave, racks of bottles bore first names and birthdates. For as long as anyone can remember, the Frescobaldis have set aside wine at the birth of each child: five hundred bottles for sons and one hundred for daughters.

We approached the mill, and I loaded fresh film in anticipation. Inside its creaking doors, I found only workaday modern machinery, a familiar red Alfa Laval grinder and spinner that churned out oil, warmed to the range of 30 degrees, like the better cooperatives in Jaén. Designed in Sweden and made in Italy, it was fine equipment. But somehow, it did not fit.

For all the harking back to old lore, oil making is entering the twenty-first century. Over acres of land, the Frescobaldis had cleared away tired, frost-crippled trees to try a popular new approach championed by Giuseppe Fontanazza at the Olive Institute in Perugia. Instead of the traditional twelve trees per acre, they planted two hundred. Each was pruned in a style called *monocono*, a Christmas-tree-like cone, to allow hydraulic vibrators to harvest its olives in half a minute. The new equipment cut pressing time to a fraction of what it was. Olives were a crop, harvested as mechanically as wheat.

But old ways remained. From Nipozzano I went to the Antinori estate at Badia a Passignano, a former abbey set in groves of old trees. At lunch, there was that oil from the night before, and I poured it in torrents on fresh pecorino cheese. After the grappa, ready for anything, I approached the mill with trepidation. It was a gem, an old-style tower press with woven *fiscoli* mats. In the final stage, a small vertical centrifuge spun off Laudemio oil at gentle speeds, and this was then filtered and sent off to market. But batches of the private stock were left to settle naturally in old clay jars. In a stone-walled backroom, under heavy oak beams, I lifted a round wooden lid and breathed deeply until my head swam.

After their first few years, Laudemio producers had mixed reactions to their consortium. It had a certain *succès d'estime* in America and Europe. Some people loved it; others dismissed it as a gimmick. Without getting into numbers, Tesi declared the venture profitable. But Bona Frescobaldi acknowledged some disappointment. Not enough people were prepared to spend so heavily on olive oil, however good. She held out hope for the future. An Antinori executive who preferred to remain nameless confessed the family had reservations about staying in the group.

Smaller producers had found a flaw. Although research and promotion were handled in common, distribution was left to each estate. Since Frescobaldi had a sophisticated worldwide network for its wines, it was far and away the dominant label. Some of the disadvantaged estates sold the same oil in different bottles, at much lower prices.

In the end, that cut-glass bottle is like the white dot on a Dunhill pipe. It signifies quality. But the difference is in the brier, not the dot. Others offer comparable, frequently better, oils for a lower price. Producers like Lapo Mazzei, whose Castello di Fonterutoli estate has been growing olives for eight centuries, take a totally different approach.

On a crisp, clear day, Jeannette and I headed to Mazzei's domain on a winding back road south of Florence, through the stone villages and flowered fields of Chianti. The contrast with Andalusia was striking. There, the scenery comes in layers: red earth below, sun above, and nothing but olives in between. In Tuscany, olives are the centerpiece of a colorful tableau edged in lavender, lemons, and all the shades of roses.

In those historic hills it is clear why poets and painters came to stay, sinking roots in a place where so much past had shaped a priceless present. Like the gentle loops of the Seine above Paris, Tuscany has its own indefinable light, reflected off walls in faded pastels. Life is a continuum; even last week's bread is the basis of a fresh, hearty soup for tomorrow.

I found Mazzei in the sun-splashed library of his old farmhouse-palazzo, on a fortified hill with a panoramic view of Siena on the horizon. The latest in an unbroken line running back to 1170, he is no recent convert to wine and oil. His family has been world-wise ever since Filippo Mazzei went to Monticello to plant Jefferson's vines and got caught up in the American Revolution. A past president of the Chianti Classico association, Mazzei is a smooth politician. He is a down-to-earth man of means, rumpled in the manner of an English lord who prefers the country to

London, who defines his life by the quality of whatever carries his name.

"Oil needs the hand of man," he said, echoing the phrase Bona Frescobaldi used with respect to wine but sounding more like the Núñez de Prados. Mazzei opposes the idea of Laudemio and similar proposals to turn out an oil that bears the black cock of Chianti Classico. Or worse, a regional government plan to market a generic Tuscan oil.

"The oils are not equal," Mazzei told me as we sat down to a lunch in which his own green-gold nectar figured prominently. "It is very difficult to make rules that are the same everywhere. True, there is a certain marketing value to a common label, a sex appeal, but you cannot do it in a region so big."

His wife, Carla, broke in with some expert testimony. "I own some trees outside of Florence, and my oil is nothing," she said. "Just over the hill, the oil is wonderful. That's how it is with olives."

Within the extranational European Union, size has economic advantages. Large companies are better situated to lobby and apply leverage. The Brussels bureaucracy gives export subsidies only to producers who sell at least one hundred tons, leaving out many small independents. A growing number of minor regulations favor mass production. But quality is another matter. With a product that requires personal attention to detail and flexibility to accommodate the whims of nature, small is usually better.

Like most Tuscans, Mazzei relies heavily on the stalwart frantoio, or frantioano, a self-fertilizing variety that yields high-quality oil at fairly constant levels from year to year. His frantoio is spiced with the hardy moraiolo and the more finicky leccino. Both need other varieties, like pendolino, for pollination. Masterful oil requires picking at the perfect time and selecting the right percentages of each variety.

Mazzei can afford to take the high road on his own. He produces only five thousand liters a year, at a substantial cost. "For us, there is no profit at all in olive oil," he said. With a slight smile, he added, "Now, wine, that is financially interesting."

Nonetheless, Fonterutoli gets around. Later, I found Mazzei's oil on the tables of Stromboli's—at the Disneyland Hotel.

For committed oil buyers, the trick is to find the Mazzeis. His oil is scattered around specialty stores in the United States and Europe. But a lovely, heavy volume brought out each year by the Tuscan government, *Vino & Olio in Toscana*, displays page after page of labels that turn up in the most unlikely places.

"In Tuscany, there are very good and varied oils, famous all over the world," Mazzei said. "In the north, in Lucca, as far as La Spezia. In central Tuscany, Florence and Siena, and in the south. It is much better when an oil is attached to a name."

For years now, he said, America has known about oils. More and more, people understand the differences in what is available between quality estate-bottled oil and the mass-produced stuff that is blended from bulk imports. "In New York, I heard people saying, 'Buy your oil where the Italians buy theirs: Spain.'"

I asked whether the Mafia still made an impact on the olive-oil trade. "We've broken that with the estate brands," Mazzei said. "Up here, we have no contact with the Sicilians. They control much, much less."

We turned our attention to Carla Mazzei's lunch, which was plain and wonderful. With the grilled veal, she drizzled Fonterutoli liberally on a plate of steaming white beans. The beans would have been enough. For serious oil tasting, nothing beats warm *fagioli*, for the texture as well as the flavor. The salad was dressed in oil with a dash of lemon and salt.

"Our favorite dishes are the simplest," she said. "Tuscans love *bruschetta*: garlic and oil on grilled bread. It is better without the garlic. Just salt or pepper. Around here, we call it *fettunta*. Take oil, pepper, and salt, stir it with a fork, and dip in the bread. In the summer, we take a piece of fresh bread with a tomato slice, crushed basil, and pecorino cheese, or parmigiana or mozzarella."

Although Italians know olive oil is healthy, Carla said, cholesterol is seldom foremost in their thoughts. People love raw egg and olive oil. A regional favorite, *bistecca a la fiorentina*, is nothing more than a generous cut of grilled Tuscan beef rubbed in oil and salt.

On the way back, we stopped so Jeannette could duck into a café for another bottle of Lapo's oil. A tableful of English tourists watched her pay 18,500 lira, $15, for a half liter.

"That sounds like a lot of money," someone said. The Laudemio price tag would have stopped their hearts. "What do you do, put it on salads?"

"That, and on bread, and beans," Jeannette replied.

"Bread?"

"Yes, like butter."

About then, the waiter arrived with an open bottle. He asked if they'd like a taste. I suspect we left behind a new table of olive converts.

Giannozzo Pucci is another oil-making Florentine noble whose roots reach down through seven centuries. His uncle Emilio's sportswear, in bold colors and geometric patterns, made the name famous far beyond Tuscany. But you would never hear that from him. He drives a rattling old Renault 4L —the little wagon with the gearshift you punch in and out of the dash— and trims his bushy gray beard only on formal occasions. He is an aesthete who can chuckle at himself, a purist in every sense. When not otherwise occupied with fighting ecological battles, he prunes his olive trees according to cycles of the moon.

Other Puccis made fortunes, but Giannozzo prefers a simple life. He writes a little and edits some, living poor on a rich piece of land that was part of a fortified village before the Medicis drew outlying citizens into Florence. He farms with passion, doing what he can to avoid the stultifying strictures of Tuscan tradition and the chemical-laced practices of big agribusiness.

"One season I decided to try out the chaos theory," he told me with a laugh. "I got sick of seeing all those long, even rows of wheat, exactly the same height. So I mixed all sorts of grain seeds and scattered them at random. It was so beautiful: all different colors, shapes. Some were short, some tall." He chuckled again. "I loved it."

Although his sister is an old pal, I had never met Giannozzo. When I phoned, he suggested we meet that evening, which was generous. That same night he had to decide whether to give up his treasured seat on the Florence City Council or make a compromise with himself.

Under new rules for the approaching election, all candidates had to declare themselves as left or right. Giannozzo, a Green, was neither. His sympathies leaned to the left, but he saw his value as belonging to no extreme. He was not a politician; he simply wanted clean air and water, a lively cultural climate, fair social programs, and livable conditions in a city that was meant for another time.

He had spent his life waging battles against vested interests. He won some, protecting the treasured old city's character with ordinances on zoning and lighting. Others were futile and exasperating. But after years on the forty-eight-member council, Giannozzo found that Florence was not changing. He was. At fifty-two, he badly wanted time to look after things that were important to him, such as his olives. Early in the morning after our dinner, he decided: the olives had won.

While Florentine politics boiled on as they always had, Giannozzo would muse and devise, explore and travel. For one thing, he would work on an idea inspired by Leonardo's oil press, now on display under the walls of Vinci. It was masterfully simple, a better way to make oil Ekron-style. Like so many believers, he attaches a mystical significance to an eternal tree that has symbolized everything rich in the Mediterranean for thousands of years. Extracting the oil it produces, he believes, ought not to exceed a natural pace.

"New is not necessarily better," Giannozzo explained. "Modern processes mean you can make oil in less time, but they don't mean better oil. My dream is to build a press that requires no more power than a single person can provide." Instead of putting the mash on *fiscoli*, he wants to use olive pits to squeeze out oil. Olives would be pressed by a hand-operated hydraulic pump. With no extra water added, he would only have to decant the olives' natural water. The Pucci press would hardly outproduce Bertolli, but that did not seem to bother him very much.

Giannozzo came by our hotel the following morning to deliver a bottle of oil. Not surprisingly, it was wonderful: cloudy but clean-tasting, with an aroma of olives and a hint of artichoke. Unlike other Tuscan oils, it came in a reused supermarket bottle, with an old label half torn off and a tin screw-on cap. And, unlike other oils, it could not be bought at any price.

In Italy, of course, the gulf between olives and politics is not wide. A million Italians grow olives; several hundred thousand enterprises handle them. About 250 brand names are found scattered around the country. The noble fruit moves Italians deeply. Early in 1996, a center-left alliance won national elections, the first time the left wing would govern Italy since it became a republic after World War II. Former Communists joined cen-trists to oppose what they called corrupt conservative elitism. Their coalition was called the Olive Tree.

Romano Prodi, who was tapped as prime minister, explained why they chose the name: "The olive is strong-rooted, resistant to bad weather, but it is soft and beautiful, the symbol of hard work and of Italy."

The bulk of Italian oil comes from redoubtable coratinas and other olives in the large region of Puglia, which stretches into the Italian boot heel. But the choice is dazzling. With this is mind, a regional booster group called Toscana Promuove organized Oleum, which was first held in March but now takes place in December. It is an olive trade fair in a city that has little trouble attracting visitors. Jaén's biennial Expoliva and the other Spanish fairs are bigger shows. Each attracts more buyers, producers, specialists, and purveyors from odd corners of the world. For many people, however, Oleum is a chance to hunt for the best oils Italy can offer.

Each year, the cognoscenti hurry to try Sardinian oil. It is not as green as the leafy Tuscan, or as sweet as buttery Ligurian, and it is lighter than oils from neighboring Sicily and the south. It has a whiff of fruity nuances all its own. Produced in small quantities, it is often hard to find. An equally prized rarity is oil from Lago di Garda in the north. A loyal band likes oil from the taggiasca, similar to the French cailletier, picked ripe around the Ligurian city of Imperia.

There are always surprises. By reputation, Sicilian oil is yellow and heavy, without much bite. But Olio Verde, made by Gianfranco Becchina from the nocellara olive on Sicily's south coast, is exactly the opposite. Deep green, lusciously unfiltered, it was terrific drizzled over raw scallops as the first course of a memorable meal.

The best oil I tasted was from a little patch of Umbria, produced by a clean-cut man of thirty-six in embroidered Levi's who talked a lot like Giannozzo Pucci.

"I've always loved olive oil," explained Ettore Donati Guerrieri. "When I first saw it, I wanted to take a bath in it. I finished law school at twenty-five and decided I didn't want to be a lawyer. Here I was, living in this beautiful place, so why not just stay there and make olive oil?"

With freckles, reddish curls, and perfect English, Ettore seems more Scot than Italian. He has a warm, easy laugh, an obvious low tolerance for pomp, and a lot of energy.

He has 4,200 trees. In southern Italy, where some sprawling warm-weather varieties can produce several hundred kilos, he would be rich. But his, severely pruned like many in the north, average ten to twenty. Along with a mix of milder local varieties, he grows leccinos for an added kick.

When he found that six thousand bottles a year was not enough for survival, Ettore decided to add a new dimension of *agriturismo*. For a reasonable price, tourists can live at his place at Castello di Mongiovino and watch him grow olives.

"This tourism thing better work," he said hopefully. "We can barely produce enough oil. A Swiss buyer was just here, insisting that I sell him my whole stock every year. I told him I could manage maybe two thousand bottles." The rest, he said, goes to faithful customers and various retail outlets, where it sells for less than ten dollars a liter.

His oil is pressed at a mill devised by an old man he calls Archimedes. A system of hammers mashes the olives and pits, and the paste is stacked in a classic tower. Rather than using mats, the layers of fresh mash are separated by drier, harder layers of pomace, *sansa* in Italian, from the previous pressing. The result is along the lines of Giannozzo's intention: olives press olives, with nothing unnatural in the process.

"We press at five hundred atmospheres, so it squeezes out all the healthy polyphenols, with lots of vitamin E," he said. (An atmosphere is the normal pressure of air at sea level.) "The oil is really green, fresh and dense, almost like cream. When it comes out, its acidity is .25, which is as low as it ever really gets, despite what anyone says."

At the end of the process, a slow centrifuge spins off the water and the fresh oil settles for a while. I asked Ettore whether he had considered old-style decantation and explained the elaborate Núñez de Prado method in Baena.

"*Truffa*," Ettore snorted. The word means trick. Also swindle. "That's the biggest *truffa* . . . to impress gullible people." He was waving his arms around, voice rising. "You have to get the water out of there as fast as possible. Water washes the oil. It ruins it."

Ettore was no more impressed with the *flor de aceite* method, which allows oil to drip naturally from mash. "You have to squeeze the olives to release what is in them," he insisted.

But Núñez de Prado oil did not taste ruined to me; its pungent olivy bouquet, peculiar to the picual, emerged perfectly intact. Ettore's oil was also excellent. The idea, both would agree, is to extract juice from olives and separate the vegetable water from the oil without adding anything extra, or disturbing molecules with heat or violence.

"The thing is simple," Ettore concluded, softening on the subject. "If olive oil is good, you know it, however it is made. When you taste that supermarket stuff, you feel it here." He stroked his chin and made a face. "It stays with you. Yuck. Olive oil has to be made with love. If you love your oil, it will be good. If you don't, it cannot be good."

As we said goodbye, Ettore tucked a bottle under my arm. Classically shaped and very dark, it looked like red wine. The label bore a line drawing of his medieval village in the hills near Perugia. "This is wonderful with everything except fish," he cautioned. "For that, you'll want something heavier, not so green, like a Ligurian."

This parting remark was the crux of it all. Every good oil is different; each has its strengths and weaknesses, depending on the culinary context in which it is eaten. Around the Mediterranean, each culture's cuisine is shaped by the local oils. Some dishes demand a specific taste. Others are spoiled if the oil's flavor is wrong.

Back home, I rubbed a fat, fresh sea bass with my own golden oil, from black Provence olives picked in December, and grilled it over olive-wood coals. With splashes of Antinori, steamed broccoli all but leaped off the plate. The next morning breakfast was simpler: Baena on toast.

On the Oleum party circuit you whisk from palace to palace, sipping fine red wine while chatting with the marchese who made it. The leitmotif of every conversation, even if about Botticelli or old silver, is olives. On my last night, under a mural-sized Tintoretto, by a wall of hand-bound leather volumes that were bestsellers before anyone wrote the Bill of Rights, I exchanged small talk.

A Turkish refinery technician told me how the worst sort of inedible olive oils are processed to make "pure olive oil." Activated earth filters, scalding steam, and chemicals remove nasty odors and murky colors. Caustic soda can cut the acidity from 5 or 6 percent to below 1 percent. That is the same level as extra-virgin oil, I remarked. The Turk laughed. "Right, except that there is no taste, no color, no body, nothing," he said. A splash of better oil makes it salable if not desirable.

A Tunisian oil merchant, proud of his olives on the Mediterranean island of Jerba, complained that Italian buyers insisted on blending his product anonymously into oils marked *Made in Italy.* Spanish growers at least had price supports and free movement within the European Union. Tunisians and Turks were out of luck. "I try to put in the contract that they have to acknowledge Tunisian origin," he said, "but no deal. They won't do it."

I spent a long time with an Australian named Joe Grilli, whose father, Primo, had emigrated from Italy. He was large, red-cheeked, and friendly, but bluff and blunt in twangy Italian. The Grillis made wine in Adelaide on hills dotted with hundred-year-old olive trees. A few years ago, Joe and his brothers decided to make oil.

"I'm not saying it's the best in the world," he said with an engaging shrug. "It's our little effort." Later, he gave me a dark, thin, long-necked bottle, elegant with a hand-made label. It was very good oil, ripe yet still a little peppery at the back of the throat.

I asked Joe what he thought about the Tuscan high greens and Laudemio in particular. He glanced furtively at the surroundings; this was no place for blasphemy. "Well, it is very good, but a little sharp for my taste . . ." he began. After a moment, the Australian overcame him. "Look," he said, belting out a laugh, "if you want chlorophyll, chew a leaf."

I later spoke to Giuseppe Grappolini, the president of MICO, the Movimento Internazionale per la Cultura dell'Olio di Oliva, which has members from Wisconsin to New Zealand. MICO is a club for olive industry notables, knowledgeable amateurs, and diehard defenders of the faith. It runs seminars for professionals who want to perfect their processes, and it lobbies for olive-friendly legislation and a tougher fight against fraud. MICO also seeks to help seed-oil eaters see the light.

"Our purpose is to win people over to olive oil by helping them to appreciate it," Grappolini explained. "Once they know its benefits, its culture, its nuances, it will be a part of their lives." His own claim to fame is an invention: a stainless-steel tasting cup with a hinged lid to capture aroma.

A star of Oleum that year was a dark-haired fireball named Alissa Mattei, whose laugh is husky and lively eyes are lined in deep blue. She'd have done well singing Gypsy torch songs, but instead she held the audience spellbound with a sober treatise on big oil. For twenty years she had directed quality control and blending at Carapelli, now the Italian branch of Ferruzzi's subsidiary, Medeol, which is the largest seller of olive oil in the world.

Carapelli was founded in 1893, but the family sold out in 1989. It is now a component of an empire that includes Koipe and Carbonell (Elosua) in Spain and LeSieur in France. The corporate ladder leads to Ferruzzi in Milan: Medeol is a branch of Eridania Beghin-Say in Paris, which is controlled by Ferruzzi's Montedison SpA. The group's olive-oil sales, approaching 200 million liters a year, make up about 16 percent of the world market. Unilever comes second, with a 14 percent share.

With their commanding positions, big conglomerates are known to play hardball. Spanish legislators charged in parliament that mysterious money changed hands to persuade Spain to allow the sale of its leading olive-oil packagers. But up close, Mattei's lab at the Carapelli plant at the edge of Florence still looks like part of a friendly little family business.

Early one morning I went to visit. Alissa bustled out from among the bottles and beakers, her gold necklaces clinking. Rather than a lab coat, she wore a nicely cut jacket. As we walked around, she peered at computer screens and paused occasionally to taste oil, with a slightly impolite slurping sound. Her assistants were hard at work testing samples for chemical content, purity, and the less tangible organoleptic properties. It was buying time, and she was busy.

Alissa's job was to find the right batches from around the Mediterranean to make up Carapelli's dozen different oils. Plain olive oil, refined and artificially colored, is not so difficult. The top-of-the-line extra virgins can be real killers. Consumers expect some year-to-year variation from estate oils which use only their own olives. Not, however, from the big brands.

"Many people find the oil they like, and that is what they want," Alissa said. "It must taste exactly the same, look exactly the same, and feel exactly the same, whenever they buy it." They may not know whether the oil came originally from Italy, Tunisia, or Zimbabwe, she acknowledged. But they know what they can expect in terms of taste, price, and quality.

International standards differentiate between virgin and extra-virgin oil, but hardly anyone packages the former. "Virgin is a very confused concept," Alissa said. Oil from healthy olives milled without delay usually rates below the 1 percent acidity ceiling for extra virgin. If not, companies prefer to blend it with plain olive oil than market yet another grade.

As we talked, Alissa stopped to test a batch of Greek oil to be certain it matched the first samples which preceded the main shipment. "You have to watch everything all the time," she said. Thick logbooks and colorful computer constructs keep track of chemical analyses. Technicians measure the tocopherols and polyphenols, antioxidant compounds which lower blood pressure and possibly fight cancer by reducing free radicals in the body.

Shelf after shelf in the lab was crammed with bottles of oils on the market. I asked Alissa if she tested the competition. "All the time," she said. Did she find irregularities? "Yes, often we come across things that are, well, strange."

Sometimes oil marked extra virgin is nowhere near virginity. Often, seed oil is mixed in. Occasionally, technicians find contaminants. The rule of thumb, Alissa said, is that large companies with high legal exposure and a reputation to maintain are more likely to police their own quality. Many others are less careful. But anyone can inadvertently mix in a batch of inferior oil, or cheat shamelessly. She named no names, but more anomalies come from southern Italy than anywhere else.

Among the olive elite, it is generally heresy to say a kind word about supermarket oil. True, it is mass-produced and without surprise. There is a certain spiritual kick in buying from, say, a Lapo Mazzei that is lost when the seller is a faceless board of directors with huge fleets of tanker trucks. Still, if buyers and blenders are good at their jobs, a big company can make respectable oil at a reasonable cost.

The big companies' economic advantage, which can be a culinary disadvantage, is that they stay close to mass-market trends. If they perceive demand, they will respond. Most are careful not to risk one extreme or another in terms of taste. Among the lineup of Carapelli oils on display,

I noticed a "lite" olive oil sold widely in the United States. Most big brands now offer Americans a light oil.

This, Alissa explained, was simply refined olive oil with less extra virgin added so that it had a clearer color. The calories are the same as any olive oil, about 125 per tablespoon. It is cheaper to make. It is inferior, with a higher percentage of chemically processed, deodorized oil. And because it is a specialty item, its retail price is higher.

Heading northwest from Florence, down from the hill country and toward the Ligurian Sea, you reach the dramatic walled city of Lucca. From up high it is a thrilling sight, densely packed red-tile roofs and stone spires within high, thick ramparts: "that compact and admirable little city" of Henry James's diary. Only a few gates lead into the narrow cobbled streets. The hills around are vibrant with color. James loved the city, "with its wide garden-land, its ancient appanage or hereditary domain, teeming and blooming with everything that is good and pleasant for man."

Lucca is closer to the heart of Tuscan olive country than its ancient rival, Florence. It was an Etruscan capital until the Romans absorbed the region in 91 B.C. Under Charlemagne, in the eighth century, the court of Lucca ruled all of Tuscany. Lucca rivaled Florence in the rich silk trade until 1284, when Genoa's navy defeated Pisa, Lucca's nearby ally. After that, Florence controlled Tuscany. Eclipsed, Lucca turned its attention to its trees, vines, and fields.

Most Chianti winemakers began only recently to sell the oil their families have made for centuries. The Lucchese, olivewise, seem to have colonized the world. Bertolli and Berio, by far the dominant names in American supermarkets, are both based in Lucca. At the other extreme, delicate estate-bottled oils come from around the old city. Lucca olives grow at lower altitudes, close to the sea, which softens their oil. Frantoios are balanced by milder pendolinos. The heavier oils of Liguria are too sweet for people who like their oil to bite back. High Tuscan is a taste some never acquire. Lucca oil is a pleasant stop halfway in between.

I had mentioned Lucca to my friend Hugues, a painter who was once married to Giannozzo Pucci's sister. "La Diamantina," he said, slapping the table for emphasis. "You must meet La Diamantina." She was, he said, a charming and quirky baroness, devoted to her olives. Despite all the expense and bother, she maintained one of the last old-style *fattorie*, producing farms, in the area. "You will love her."

Hugues was right. At least partially. Hugues remembered she had once told him: *"Sono brutta, vecchia ma molto simpatica."* I am ugly, old, but very nice. If she said it, only the *simpatica* part was accurate. Diamantina Scuola Camerino had the handsome beauty of a woman with little patience for externals. Without makeup, hair simply cut, she let character and breeding speak for themselves. She wore a tweed jacket and sensible shoes. After a moment's conversation, it was clear that she was more concerned with the appearance of her trees.

A great pale yellow house looms over Forci, her estate. Seven arches on cut-stone columns form a long, open verandah, which overlooks a plant lover's paradise. A blooming oleander reaches almost to the second-story roof beams, and bright flowers spill down a slope out front. The old library includes *Viaggio in Toscana*, a German traveler's account of the estate nearly three hundred years ago. A freeze in 1709 killed most of the trees. Within a decade they were thriving again.

Diamantina is in a curious position, a homespun farm girl who also belongs to one of the fancier families in a country that takes status seriously. Her name comes from a property her father owned, across from a fabled castle known as La Diamante, the Diamond. But he preferred Forci, with its lovely remote hills and olive groves. Forci still gets a lot of visitors. Once, a president of Italy dropped by to see the press. Although Diamantina spends much of her time in Rome, she hurries home whenever she can.

When I mentioned the theory that the world was divided into olive people and butter people, she laughed. "I cultivate olives, but I adore butter," Diamantina said. "There's room for both." With lunch, she offered some of the best oil in Lucca. It was fruity but not green, neither thin nor greasy, wonderfully mellow but with a subtle, sharp bite just when you thought you were finished tasting it. She did not claim a lot of credit for what was done by nature, a skilled foreman, and a very clean press.

We picked up Armando Scaramucci, the foreman, and headed for the mill. The original press was a masterwork, built into an old stone wall. Beautifully shaped dark-wood beams framed the top of it. Pressure came from two enormous wooden screws tightened by a pole worn smooth by ten generations of hands. The date, 1759, and Forci's star-shaped emblem were carved into the lintel. These days, Diamantina uses a slightly newer version.

"I used to love to watch them make oil when I was a girl," she said. The estate label is still the same pen-and-ink line drawing she did for her father when she was a kid: the press, along with its granite crushers and wooden gears. "It was only a father's love that made him use it, I think," she said. "But we like it now."

Armando poured some oil for a serious sampling. The Forci style was like ours in Provence: teaspoon, hunk of bread, or, *à la rigueur*, an index finger. With some merriment, however, the two of them showed us how a developing school of oil tasting does it properly. Later, Armando presented me with an official kit, complete with a 24-page illustrated manual, two of MICO president Giuseppe Grappolini's stainless-steel vessels, and a tiny towel.

"The tasters take themselves very seriously," Diamantina said. "You are supposed to bring the oil through your teeth with great force, smack your lips loudly, and spit out the oil hard so you clear your palate." She grimaced at the thought.

One time, she remembers, forty amateur tasters visited Forci. She did not know about it beforehand. "I walked by the mill and heard this horrible sucking sound," she said. "You can imagine what I thought. Here were all these people, intent as if they were in a trance, making weird gargling noises, spitting on my floor, and staring downward. It was like some Satanic rite."

The tasting business can get a bit pretentious. One useful little British book on olive oil offers a list of necessities for a proper tasting: white bread cubes, apple slices, paper napkins, fizzy water. Or, perhaps, party hats, false noses, Dvořák tapes. An Italian manual I saw counsels against wearing aftershave, perfume, or smoke-scented clothing—or using a flavored toothpaste.

Some aficionados like to debate the nuances they detect in the oil's bouquet. The International Olive Oil Council has compiled a lexicon of terms. First there are styles: aggressive, assertive, or pungent; bitter; delicate or gentle; fresh; fruity; green; harmonious or balanced; rustic or earthy;

spicy; strong; sweet. Then there are aromas and flavors: apple, banana, lychee, melon, pear, ripe olive, tomato, eucalyptus, grass, flower, green leaf, hay, leafy, mint, herb, violet, avocado, almond, Brazil nut, walnut. And chocolate.

The defects require less subtlety to imagine. Rancid means the oil has oxidized with air. Soapy, fatty, or greasy are straightforward enough; olive-fly larvae can cause this. Earthy, too, is what it sounds like; someone probably collected olives from the ground and didn't wash them. Flat means the oil is simply not up to it, perhaps because of too much water in the olive.

Basically, tasting is only tasting. The main thing, Grappolini says, is to avoid any competing oils or aromas. He keeps his palate clean with only mineral water and unsalted bread. Oil is best warmed by hand in a snifter, like cognac, to release volatile flavor compounds. Aroma is crucial. Alissa Mattei says she knows half of what she needs to about an oil before taking a sip. It should get to the taste buds toward the back of the tongue and stay there long enough to deliver its finish.

Official panel-taste score sheets for determining organoleptic points mention only apples, ripe or green olives, and "other ripe fruits." "Green" is gauged as leafy or grassy. Tasters note whether an oil is bitter, peppery, or sweet. They look for hints of almond or hay. Scores plummet if the oil is crude, fusty or moldy, metallic, rancid, or tainted with waste water.

The eventual verdict, of course, is valid only for the batch tasted. Oils vary by the year. Their flavor often shifts during the first months. Within two years, they are likely to go rancid, although not necessarily. And in the end, oil is seldom consumed by the teaspoon or shot glass. Olive oil takes on the flavor of the foods it accompanies.

A generation ago, Forci was tended by twenty-eight *contadini*, Italian peasants of an old system linked loosely to serfdom. Families lived free on the property and worked a plot of their own, but they devoted themselves to the master's estate. There were no schedules and few statutes, only work to be done according to needs and the seasons.

These days, Italy's rigorous labor laws apply to agriculture as well as industry: minimum wages, social charges, strict hours, vacations, retirements, sick leave. Even rich estate owners can afford only a few hired hands. "It is fair that people should have a short work week and vacations," Diamantina said, "but the trees don't take time off." Only Armando lives the job, day and night, and he seems to love it.

During his infrequent breaks, Armando paints landscapes and sea-scapes in vibrant oils. More often, his artistic urges go into shaping Forci's old trees. That afternoon I walked the groves with him, an acolyte in the wake of a priest. After my visit to Palestine, this was the other extreme. Here, the unimaginable was available: experimental chemicals, newfangled gear, radical techniques. The problem was to find the right balance between old and new. Armando managed to be a daring conservative.

We chatted about iron sulphate, malathion, and urea, substances only an oliveman could love. Extra iron in the soil cures chlorosis, which yel-lows the leaves. Malathion kills pests. Olive trees crave nitrogen, even from cow urine. But too much of an additive is worse than none at all. Pesti-cides, especially, are easily abused. The wrong thing at the wrong time murders the bees. Systematic spraying builds immunities and throws off nature.

Lucca is good oil country, Armando said, because it is far enough from the coast to be spared from olive fly. Most pests attack the trunk and branches. An alert grower can treat them without trouble. But the deadly dacus leaves its larvae in the fruit. Damaged olives drop off, or they flavor the oil with an earthy—read wormy—taste. Throughout the Mediterra-nean the situation is similar. Trees by the shore tend to be bushy and tall, nourished by moisture in the air. The olives are plump with water, and the flies love them. Trees grown inland, higher up, are tougher, and their oil is usually better.

Armando prunes the older trees hard every two years, renewing the wood to keep production high. He displayed these sculptures with an ar-tist's pride. He also showed me a new grove planted in the new Fontanazza fashion, like Frescobaldi's, each separated by only three meters. He, too, was thinking about harvesting by vibrator. The most popular was a tractor-hauled unit which clutched the trunk with a mechanical paw and shook it hard. A more elaborate one was a high, wide, U-shaped vehicle that straddled bushy little trees that looked like sheep waiting to be sheared. Other devices could prune mechanically.

Armando admired Fontanazza and knew him well. His method, initiated in the 1980s, was catching on. In a country short of farmhands, it was supposed to cut a year's harvesting and pruning time by a third, to fifty hours an acre. And where soil and terrain allowed, as many as five hundred trees could be crowded into one acre. Drip irrigation regulated moisture. But something seemed frightening about this new trend. Mechanized olives? In a country that produced varieties with names like gentile di Larino and Sant'Agostino, Fontanazza's prize breed is "I-77 Clone."

I told Armando about my struggle back home. Centuries-old trees were weak from long neglect, but the trunks were beautiful. Cut them down and start again from the stump, he said. After the 1985 freeze, Lucca growers faced the same dilemma. Everyone who cut immediately now has healthy, full trees. Those who waited found that they eventually had to haul out the chain saw. I preferred to nurse along my living historical monuments.

Finally, we talked about pressing. A perfectionist, he had carefully examined the options. He liked the traditional presses, but they were too hard to keep clean. They had to be operated without stopping or the *fiscoli* could turn rancid. Stone crushing exposed the mash to oxidation for too much time. Older centrifugal systems were too brutal, with a risk of "washing" the oil or damaging it with heat. A third process, known as sinolea, separated oil droplets from vegetable water with a grid of moving filters like stainless-steel fingers; oil clung to the metal, but water did not. This was expensive.

Armando saw the most promise in new continuous systems on the market. Pieralisi, Alfa Laval, and Rapanelli, among others, had designed centrifuges that used only the olives' natural water, unheated. They spun at gentle speeds. These solved a growing ecological problem: what to do with the lakes of greasy black waste liquid after milling. They also made good oil, unaltered by high temperatures and unsullied by excess water.

Even with steady improvements in the process, however, the essential element remained unchanged. A clumsy miller could make inferior oil with the best equipment. But no one, no matter what he might do, could make good oil from bad olives.

In the end, Armando concluded, so many factors come into play that you can never be certain what each year will bring. Estate-bottled oils will always vary in taste from year to year. The only oil with a uniform taste comes from producers who blend it. And then you can never be sure what you are getting.

"Laws control the manipulation of oil, but . . ." He made a distinctly Italian gesture meaning, Don't count on it. An old hand at the business, he knew all about the scoundrels he described variously as crooks and mafiosi. "Old or bad oil might be chemically reconstituted and mixed in. It might be cut with vegetable oil, or who knows what. There are so many ways to manipulate it. Buying oil is like going to the doctor. You must have faith in whom you're dealing with."

Branch Tips

Sicilian Swordfish in Foil Packets

Nancy Harmon Jenkins, author of *The Mediterranean Diet Cookbook*, likes this recipe for entertaining, especially because there is no last-minute mess.

½ pound swordfish or tuna steak, about 1½ inches thick
A little flour for dusting the steak
1–2 tablespoons extra-virgin olive oil as needed
1 small onion, thinly sliced
½ garlic clove, minced

¼ cup chopped pitted green olives
2- inch strip of lemon zest, finely slivered
1–2 teaspoons tomato puree, diluted with 1 cup dry white wine
Salt and freshly ground black pepper to taste

Dust each side of the fish lightly with flour, shaking off the excess. Heat 1 tablespoon of the oil in a sauté pan over medium-high heat and sauté the fish quickly just until it is golden on each side. Remove and set aside. Lower the heat and in the oil left in the pan sauté the onion and garlic until they are soft, adding more oil if necessary. Then add the olives and lemon zest, and cook for 2 or 3 minutes. Add the tomato puree; then raise the heat slightly until the wine is reduced and the sauce thickens. Place the fish on a large square of aluminum foil, pour the sauce on top, and fold the ends into a loose but tightly sealed packet. Preheat the oven to 425 degrees. Put the packet on a cookie sheet or a shallow baking pan and bake 15 minutes.

Serves 2.

Adapted from *The Mediterranean Diet Cookbook*

"Since he had started off relatively helpless,
economically, since he did not believe in advertising,
relying on word of mouth, and since if the truth
be told, his olive oil was no better than his
competitors', he could not use the common strangleholds
of legitimate businessmen. He had to rely on the
force of his own personality and his reputation as a
'man of respect.' "
—From The Godfather

"I made it all up."
—Mario Puzo

7

Cosa Nostra

Back in Madrid, when a Spanish oilman heaped abuse on Italians who had cornered the American market, he shrugged and concluded: "Didn't you see *The Godfather?*" As any movie-trivia expert will recall, Don Vito Corleone was shot outside his family olive-oil business in Manhattan, the heart of his empire. Having heard so much about Mafia involvement over the years, I decided to ask Mario Puzo about his research for the novel.

"I know nothing about olives and less about the real Mafia," Puzo replied. "Do they really exist? I made it all up." He didn't, of course.

"*The Godfather* was a seminal work," Nick Pileggi told me later. Pileggi, a deep-digging reporter since his days at Associated Press, is the author of *Wise Guy*, the profile of a mobster who maneuvered among the New York crime families. "Puzo had it exactly right." He gave me a page full of phone numbers, and I began to poke around.

As it turns out, for most of this century olive oil has been as dear to the Mafia as concrete shoes. Sicilian crime families first sank their roots into America just after the Civil War, bringing Italian farm workers to Louisiana after freed black slaves left their jobs. By 1902, they had so much to protect that a hit squad murdered Joe Petrosino, a New York detective, as he nosed around Palermo.

Among their criminal activities, mafiosi imported staple foods from home for Italian neighborhoods growing quickly in big cities. There were artichoke kings, cheese barons, and, obviously enough, oil moguls. This was only logical. It was a service to the community. What decent Italian mother could last out the week without thick, yellow southern oil to pour into minestrone? It was a legal, handy cover. The Sicilians knew where to get steady supplies. And the rewards, if lucrative, were not enough to tempt any serious challengers.

After Prohibition was repealed in 1933, the business took off. "A lot of guys had all these trucks and nothing to deliver," Pileggi told me. "Some went into legal liquor distributorships. Others went into olive oil. Several families got involved. When the Profacis moved into olive oil, it was their first legitimate business."

But, I asked, didn't these men of respect lean a little heavily on foolhardy competitors and enforce monopolies within their various territories? "Well, yeah." Pileggi laughed. "If you tried to compete, you went up against some pretty tough guys. Still, that was not unlike trying to compete against the Rockefellers in oil or Carnegie in steel. It's just the free-enterprise system."

When World War II ended, demand was high. Italian restaurants needed increasing quantities of oil. Later, as pizzerias began sprouting up on every corner, the market broadened, not only for oil, but also for cheese and tomato paste. Eateries were a valuable cover for mob activity. Friends and strangers could drop in as often as they wanted without having to explain why. Anything that got discussed, or sold, in back rooms was hard for the police to monitor.

By the 1940s, Joseph Profaci had emerged as the Olive Oil King. He controlled the docks in New York and New Jersey and cut deals with other *capos* who sold oil in other territories across the country. In *Gombata*, John Cummings and Ernest Volkman described the reigning head of the Colombo crime family as "an unbelievably cheap man." They wrote:

He was a multimillionaire, with vast income from assorted rackets and his own legitimate business, which was a virtual monopoly over all olive oil sold in New York City. A family man and devout churchgoer, Profaci nevertheless had a reputation for ferocity. Once, when a local junkie stole gems from the crowned head of the Madonna in Profaci's parish church, the man known in the Mafia as "the Olive Oil King" had the thief tortured for hours, then strangled to death with a set of rosary beads.

Not exactly, but close, according to the man who probably knows best, a retired New York cop named Ralph Salerno, who has settled in Naples, Florida, after nearly fifty years of watching the mobs. The jeweled tiara was stolen twice, and Carlo Gambino helped get it back the second time. The thieving punk did meet an untimely, painful death, but Salerno did not see any rosary beads. Profaci was a patron of Regina Pacis Church in Brooklyn. When Monsignor Cioffi presented the church to the Pope, he revealed a Michelangelo-style painting on the ceiling. It featured a crowd of the faithful. Most of them were faceless, but there was a clear likeness of Joe Profaci.

"The pastor used to tell the cops and the FBI, 'Leave him alone, he's a wonderful man,'" Salerno remembers. "He'd tell us to go out and catch real criminals. That's how it was. Once a partner and I tailed a guy into a church and saw him do the Stations of the Cross. 'What a hypocritical bastard,' my partner said. 'No, he's not,' I said. 'He's religious. The other stuff is strictly business.'"

Profaci was hatchet-faced and sharp-nosed, a short man who dressed well. He was a family man, one of the few Mafia lords who had no mistresses. No one ever remembers seeing him go to a nightclub. He was proud of his membership in the Knights of Columbus. Mostly, he stayed home and ate meals laced with better olive oil than he sold.

Home was a forty-acre estate in New Jersey, once owned by Teddy Roosevelt, which was alive with his own kids, related Profacis, and friends. The farmhouse had thirty rooms in addition to a chapel with an altar carved as a replica of the one in St. Peter's Basilica in Rome.

Profaci's niece Rosalie remembers him as boisterous, with a temper, but hardly cheap. "He was a flamboyant man who smoked big cigars, drove big black Cadillacs, and did things like buy tickets to a Broadway play for us cousins," she wrote in *Mafia Marriage*. "But he didn't buy two or three or even four seats, he bought a whole row. I remember one time when, in an uncharacteristic gesture, my father acted like his brother. He walked into a room where we cousins were celebrating the New Year and threw fistfuls of dollars in the air, and laughed as he watched us dive for the bills."

In 1956 Rosalie married Salvatore (Bill) Bonnano, linking two families that had been close for decades. As a favor to the Profacis, Joseph Bonnano—Bill's father—had gone to Canada in 1926 to bring Salvatore Profaci—Rosalie's father—into the United States from Sicily.

As a young reporter, I had covered the Bonnanos' doings during the 1960s in the *Arizona Daily Star* in Tucson, where they lived most of the time, but I didn't know that "Joe Bananas" sold olive oil. With his links to Profaci, he handled distribution in parts of the Midwest, alongside his cheese business. Other crime families also imported oil. A few enforced near-monopolies in the territories they ran, while some were content to share shelf space, as long as their brand was there.

By 1960, the Olive Oil King was at war. Joseph Colombo, with the help of Carlo Gambino, put the word out that he would take better care of disgruntled wise guys in the Profaci organization. "Crazy Joey" Gallo, one of those young malcontents, came out shooting.

Gallo's hotheads were satirized in Jimmy Breslin's *The Gang That Couldn't Shoot Straight*, but Salerno was not impressed with his version. "About sixteen people were killed, and not one innocent bystander was hit," he said. "They shot pretty straight." Salerno and his friends put Gallo away for seven to fourteen years, on extortion charges. Profaci died of cancer in 1962. Then the Banana Wars broke out.

Besieged by the Gallo gang, Profaci had appealed to Bonnano for help. The Mafia Commission of family *capos* ruled that the war was an internal affair, and Bonnano stayed out. But the families were in disarray. The FBI had raided a meeting of the high command at Apalachin, New York. Joe Valachi, a workaday mobster, bargained for immunity. He referred often to La Cosa Nostra—our thing—in detailed testimony that made the term a household word. Prosecutors went all-out against Joe Bonnano.

Bill Bonnano moved into the home of Joseph Magliocco, a relative who was trying without luck to succeed Profaci. This, for some, was seen as an alliance, and Bill was suspected of helping in a failed plot to whack Carlo Gambino and Thomas Lucchese. Joe Bonnano was reported kidnapped—he resurfaced, and the facts were never made clear—and Bill disappeared for a while. Profaci's family was up for grabs. The last thing anyone worried about was olive oil.

These days the business is a little calmer. "No doubt about it," said James Fox, who retired as chief of the New York FBI office in 1994. "People are making payoffs, and some of the old families have their fingers in the business." But as other active FBI agents admit, details are scarce. "Quite frankly," one told me, "we don't pay much attention."

What the FBI's organized-crime agents call "LCN," for La Cosa Nostra, has cloaked most of its businesses with legitimacy. They still control the distribution of a lot of the food eaten in New York and elsewhere, as well as the companies that cart away what is left over. They sell concrete for most of the city's construction. But the range of racketeering, extortion, the drug trade, gambling, and prostitution is run less by Italians than by Asians and Latin Americans. And now the Russians are closing in fast. For hard-pressed law officers, occupied with murder and mayhem, olive oil is not much of a priority.

"The same conditions no longer prevail," Salerno said. "Up to the sixties, you had sizable Italian communities in the Eastern United States. No more. What was Little Italy in New York has been swallowed up by Chinatown. Some old shops remain, but you can't find an Italian above the ground floor. The largest Italian community was East Harlem. Now that is Spanish Harlem."

The Mafia has no press office, and candor about business dealings is not a mob trait. That leaves few reasonably reliable sources, so just for the hell of it, I thought I'd try my luck among retailers.

Dean & Deluca, a classy mega-deli which has expanded along the

Eastern Seaboard from lower Manhattan, is a reliable fount of good oil. Its selection is wide, from small bottles of Crete's best to five-liter cans of Núñez de Prado. Joel Dean looks after much of the operation. But olives and oil fall under his partner, Giorgio Deluca.

Nothing in particular suggested that this upstanding Italian merchant, a gentleman and gourmet, would know about Mafia involvement in the oil trade. Still, it was worth a try. Deluca had built up his food empire in a tough neighborhood, and there was little he did not know about the oil business.

Short and wiry, with gray curls and an amused squint, Deluca operates on fast forward. A man with no time to waste, he manages to be courtly and curt at the same time. I found him in a cluttered little office. On the wall, in the spot where some people like to tape a photocopy of Desiderata, he had a passage from Genghis Khan: Man's purpose in life is to plunder, kill, carry off the enemy's wealth and women, and so on. I think he meant it as a joke.

Deluca relaxed a bit on learning that I had nothing to sell. He enjoys talking about olive oil. We chatted awhile about tastes and trends. Americans are catching on fast, he said, and a lot of people pay heavily for quality. Soon it was time to let him get back to something more productive. With no better opening, I made a wild stab. In Italy, I said, people had told me that the popularity of small brands had broken the monopoly of bigger players. Before, olive oil had seemed to be largely in the hands of a few guys in Sicily.

"What do you mean?"

"Well, it seems that the general assumption is that the Mafia used to control a large part of the olive-oil imports to the United States."

His amused squint flickered. "The stuff leaks in from everywhere," Deluca said, picking up my card and studying it.

It is true, as Lapo Mazzei said in Tuscany, that the proliferation of small brands and large multinationals has broken the mob's old stranglehold. Olive oil does leak into the United States from all over the world. A few brands of heavy yellow southern Italian oil no longer dominate market shelves.

And there is the phenomenon of generational shift. Joe Bonanno turned ninety in Tucson, with no thugs at the gate. Bill and Joe Jr. have been investigated for various scams, but neither has been charged with making anyone offers they could not refuse.

Educated scions of mob families were finding it easier to acquire money with a pen than an assault rifle, U.S. News & World Report noted in 1988, in an article headlined YUPPIE MOBSTERS OPT FOR BRIEFCASES OVER GUNS; FROM ORGANIZED CRIME TO ORGANIZATION MEN. Among other cases, it reported: "The sons of the late Joseph Profaci, an olive-oil king and 30-year head of one of New York's top Cosa Nostra clans, work in the Italian foods business."

After the elder Profaci died, Joseph Colombo took over the crime family and gave it his name. He was gunned down in 1971. Salvatore Profaci, the oldest son, assumed control of his father's private holdings. And now Sal Profaci is, as they say, a reputed capo of New York's Colombo family.

For 435 days over two years in the early 1990s, Profaci and associates unwittingly made tapes for the FBI. Agents bugged dozens of organized crime figures in the Camden, New Jersey, law offices of defense attorney Salvatore J. Avena. The operation cost $517,673.

As George Anastasia reported in The Philadelphia Inquirer on October 9, 1994:

> They talked of cutting out the tongue of a young Phila-
> delphia mobster, and of burying him and two others in quick-
> dry cement. They mocked an old bookie who begged for his life
> after a package containing a dead fish and a bullet arrived at his
> door. They spoke of crushing a trash tycoon in his own com-
> pacter . . . They considered recruiting hit men from Sicily and
> New York to rub out dissidents in the Philadelphia underworld.
> La Cosa Nostra, the New York gangster [Sal Profaci] told his
> counterparts from Philadelphia and Scranton, "is a beautiful way
> of life if we respect it." Mob talk.

At one point, Sal Profaci philosophized: "My grandfather, when I was a little boy (told me) 'Salvatore, when strength and reason oppose each other, strength will win and reason becomes worthless.' " Later, he added, La Cosa Nostra was sacred.

Profaci boasted how his New Jersey company, Full-Line Foods, had muscled into wholesale pasta sales to a large supermarket chain. In summarizing this, U.S. prosecutors said they had begun extortion investigations. At another point Profaci took offense when someone mentioned a competitor was making inroads into pizza shops in the area. "I'm in the pizza shops, I'm in all the pizza shops," he noted with a clear menace, "and if I'm not there today I will be there tomorrow."

Among the best olive-oil bargains in America is Colavita extra virgin. Produced in Molise, it is mellower than Umbrian but sharper than oils from farther south. For its quality, it is surprisingly inexpensive. Colavita is run by a young Harvard-trained lawyer named Joseph Profaci.

"What can I say, he was my grandfather," Profaci said with an awkward laugh. "There are a lot of Profacis, and as far as I know, I'm related to all of them. With the kind of scrutiny we face, you can imagine how careful we are." I could. Joe, unimposing, with a gentle manner, a quick friendly smile, and a child's seat in the back of his sensible sedan, is the exact opposite of anyone's wise-guy stereotype. From every indication he seems to be, as Israelis would say, good olive oil.

During 1995 he took over active management from his father, John, giving up a fledgling law practice because he likes the oil business. He preferred to talk about Colavita, but answered all my questions with good grace.

"My grandfather took my uncle Sal into the family business," Joe said. "He was the oldest. My father was kept pretty much out of it. That's how it worked. We kids didn't know anything. I remember hearing about the Mafia in school and going to the library to try to learn something about it." He winced when I mentioned the Sal Profaci tapes. "That was a terrible embarrassment," he said, looking terribly embarrassed.

John Profaci started the business in 1978 with Enrico Colavita, an Italian oliveman, who came to America on his honeymoon hoping to find someone to distribute his oil. Through a cousin, he met John, a food broker, for lunch at the New York Athletic Club. They settled the deal with a handshake. John is still president but, semi-retired, spends most of his time promoting olive oil and Italian cooking. His reputation, like Joe's, is without a blemish.

At Colavita headquarters in Linden, New Jersey, Joe took me on a tour of his empire. The 60,000-square-foot warehouse was nearly empty. At times, sales come close to getting ahead of the olive crop. Colavita buys the oil from nearby producers and bottles it at his plant in Campobasso, Italy. He blends four varieties: gentile di Larina and Collina di Rotello from Molise, coratina from Puglia, and leccino from Abruzzo. From there the oil is shipped, crated, and ready to go, to Linden.

In sales volume, it does not approach Bertolli or Filippo Berio, but I find it fruitier, livelier, and of a lighter texture than either of those. The label is handsome, but packaging is nothing fancy. "We try hard to keep the price within reach for the average consumer," Joe said. "People seem to like it."

His father, John Profaci, takes pride in the honors given him by the Culinary Institute of America in Hyde Park, New York. He endowed the Colavita Center for Italian Cooking on the campus, where each of the institute's two thousand student chefs learn to make lives happier with olive oil. Friends say he is eager to go down in history as something other than the son of a major mobster.

"My father has worked so hard to build up this business, and to promote the image of olive oil in America," Joe put it. "He hates it when attention gets distracted to the past."

Months later, I decided to explore the Mafia connection from another angle, Italy, and Lucca seemed a good place to start.

During the last century, an enterprising Lucchese named Filippo Berio began shipping oil to Italian communities all over the New World. His business boomed, and he held an edge over others who followed his example. Dino Fontana bought the Berio name early in the 1900s, and he thrived. Fontana's main competitor was a neighbor, a man named Bertolli, who was also a friend. After work, they raced cars together on the Italian circuit. Between them, they dominated the American market.

Today, Bertolli accounts for about a third of the sales in the United

States, and Berio is close behind. Bertolli, now owned by Unilever, is merely a brand name. Berio is the only large exporter still in the hands of a single family: four grandsons of Dino Fontana. They are friends of Diamantina, and when she conveyed my interest in olives to them, Luigi and Alberto Fontana invited us to a lavish lunch.

It would be an interesting exercise. Chances were slim that a tradition-bound family which had prospered into the fourth generation at the same business would be messing around with Sicilian thugs. Berio sold good, reliable oil, at an affordable price, and I'd learn much from the Fontanas about olives in general. At the same time, these astute international merchants had to know every detail of the territory. The trick would be probing for details without creating an awkward situation for Diamantina, who was not involved in the seamier side of olive oil.

Berio's ceremonial headquarters is a Tuscan farmhouse, a country mansion of sprawling, low, pastel-washed buildings with tile roofs. The gardens are lush and rambling. A cheery fire warmed the den, its light reflecting off framed photos of Dino Fontana at the wheel of the race car that eventually killed him. A tiled inlay on the floor by the main door read: *Home of Italy's largest olive orchard.* This was no idle boast.

Out back were 45,000 two-year-old trees; along with classic Tuscan frantoios were the pendolinos that often go into luscious Lucca oil. The 200-acre parcel had been waterlogged marshland only a few years earlier. Fifty miles of underground tubing drew off the excess moisture. In a few more years, the trees would be shaped in *monocono* so that hydraulic vibrators could harvest them industrial-style.

Growing its own olives in quantity would be a change for Berio. Like that of the other big Italian producers, the bulk of Berio's oil is hauled in from Spain, Greece, and elsewhere. Each year's mix is a closely guarded secret. This, the Fontanas maintain, is no cause for shame: purists be damned, bulk does not necessarily mean an inferior product. The company's claim that its oil is the best in the world may be a little exaggerated. But it is certainly not the worst.

"The secret of good oil is in the blending," Alberto Fontana explained, echoing the point Alissa Mattei had made at Carapelli in Florence. "You have to find the right tastes, and this you learn with experience."

Alberto cuts an imposing figure, with thick black brows over sleepy eyes and a subtle smirk that suggests a great deal of self-confidence. His slang-laced English is perfect from years of running the business in New York. He now directs international marketing.

Luigi, Alberto's older brother, handles production. Gray-haired, gray-suited, wearing a tightly knotted sober silk tie, he is a textbook Italian paterfamilias. Luigi is in charge of buying and blending Berio oil. Like Alberto, he exudes the air of a man used to being taken seriously.

When I mentioned to Luigi the various artisanal pressing processes that small producers hold dear, he waved a dismissive hand. "Modern methods are the only way," he said. "Technology has been so perfected that you can produce great quantities of extra-virgin oil without losing any quality. On the contrary, it is better." He nearly had me convinced until we sat down to lunch. I doused a spicy tomato risotto with oil from a cruet on the table. Soon, I was sipping the oil from a spoon. Cloudy and golden-green, mellow with a subtle bite, fresh and full, it was one of the best oils I had ever tasted. "Oh, this," Luigi said when I asked about it. "We make only a few liters for the family." It was Lucca oil, pressed the old-fashioned way.

Over the roast veal and salad, Luigi told me of his own oil preferences. In Greece, he liked Crete the best. Crops were usually abundant, with dependable quality. The oils were similar in taste and consistency to Italian extra virgins. At the mention of Tunisia, he made a face. "Too harsh," he said, "not enough attention to the trees and harvest." And he was not impressed with most Spanish oils, especially the bulk production around Jaén. A lot of Italian packagers used Andalusian oil, he said, but the reason was price, not quality.

I told Luigi about the Spanish complaint that Italians had cornered the American market without leaving them even a look-in.

"The Spanish don't how to sell, that's their problem," he said. "They insist on sending over the oil that they like, not what the customers want. That's the important thing. If you sell oil, you have to offer something that appeals to local tastes. No one can force people to buy one thing or another. It is up to the seller to provide the best product."

After a few more bites, Luigi put down his fork. "Northern Portugal," he said. "Now there they have some wonderful oil, up by the Extremadura border with Spain. Excellent. We never manage to buy any because they don't produce enough of it."

A growing worldwide demand for good extra-virgin oil put pressure on the big companies, Luigi explained. When buyers locate the suppliers they want, they often find the price is too high to allow the profit margin decreed by headquarters. Some are forced to buy second-choice lots. Or worse. This is what separates the blenders, he concluded. Any good taster can recognize superior oil. The question is, will they pay for it?

"We can't afford to cut corners," he said. "Everyone knows our name, and we stand by a reputation that goes back generations. Multinationals or companies out for a fast profit might settle for something inferior. The Fontanas cannot."

By the time the homemade lemon gelato arrived, my soul was in conflict. Here I was, a welcome and honored guest at a family table, courtesy of a friend. On either side were Fontanas of the next generation, wholesome and eager university students who took riding lessons. But I was captivated by Luigi's nervous tic. He repeatedly tilted back his head and rolled out his lower jaw, a dead ringer for Brando as Don Vito Corleone. Alberto could have doubled as the oldest son, Santino. I was dying to ask about unrefusable offers.

Over coffee and grappa by the fireplace, Luigi began to glance at his watch. This was a workday. Casually I mentioned to Alberto that some of the Sicilian families in the United States started out by exporting olive oil. He nodded, lids half closed, waiting to see where I was headed. "Is there any Mafia influence left in the oil business?" I asked. "Not that I know of," he replied. I asked if his father had relayed any such lore from the past. "From what I know," he said, "we have never had any threats or pressure of any kind." He seemed more amused than troubled by the subject.

I said that certain old Mafia family names recurred among today's oil importers and mentioned one in particular. Was it possible that there had been a generation shift and sons had legitimized operations that had been covers for something else? I tossed out a name about which I was curious. "In the case of that particular guy," he said, "I'm not so sure there has been any change."

This was in English. Luigi had followed only sketchily, and he looked over at his brother. By way of translation, Alberto asked in rapid Italian: "It's true that [my example] is still mafiosi, no?" Luigi shrugged. Moments later, we exchanged pleasantries and I left.

From Lucca, I had to choose my next stop. One logical choice might be Sicily. I had been there before, writing about the Mafia, and sifting fact from hearsay is not easy in Sicily. Prosecutors in Palermo, busy enough trying to stay alive and confront major crimes, were not looking into olive oil. Corleone is a charming stone hilltop town of donkeys and grandmothers in black stockings, surrounded by old olives, lemon trees, and rolling fields of grain. Its economy is improved by busloads of nervous American tourists on the *Godfather* trail. Big-time mafiosi come out only for weddings and funerals, and they don't talk a lot. I went to Rome.

Talking to Italian authorities and Rome-based investigators, most of whom preferred to remain anonymous, I realized how much the picture had changed. Crime families, weakened by internal wars over the drug trade, were reorganizing to face new competition from Russian mobs. The mafiosi who export oil to the United States are no longer mainly Sicilians. Their volume is not a large part of the market. Organized crime now makes money from olives the same way independent chiselers do. For one thing, they cut good olive oil with something cheaper, and occasionally more dangerous, than what is on the label.

In his 1990 reference book, *Olive Oil*, Paul Kiritsakis of the Greek Technological Educational Institution in Thessaloniki confirms what every exporter knows:

> Olive oil commands a greater price in the international market than other vegetable oils. Consequently adulteration of olive oil with cheaper oils is a temptation. Some cases . . . are hazardous to the public's health. Oils known to be widely used . . . include: olive pomace oil, corn oil, peanut oil, cottonseed oil, sunflower oil, soybean oil and poppy seed oil. In addition, castor oil, pork fat (lard), as well as other animal fats have been occasionally used in small quantities. Adulteration with reester-

ified oils, and with industrial grade colza (rapeseed oil) has also been reported.

After an industry sampling found that many brands sold in the United States contained undeclared chemically treated olive oil and seed oil, the Food and Drug Administration set up controls. This led to a cut in fraudulent labeling, but it hardly eradicated it. Like the FBI, the FDA has its priorities, and olive oil is not at the top of its list.

In the mid-1990s a trade group called the North American Olive Oil Association began its own self-policing campaign. Samples from shelves are analyzed by the International Olive Oil Council. Offenders are reported to the FDA. This has helped, but confusion remains. Early in 1996, *The New York Times* reported that of seventy-three olive oils tested by the FDA, only 4 percent were pure olive oil. Days later, the *Times* apologized. Actually, it said, only one failed the test. Nonetheless, the article was not totally wrong: adulteration is a problem.

By far the most tragic case involved Spaniards, in 1981, when at least 402 people died. Perhaps 20,000 others suffered toxic effects. As it emerged in court eight years later, some businessmen had imported industrial rapeseed oil from France. It was colored with red aniline dye to warn that it was meant for lubricating machines. The importers removed the dye, packaged the oil in five-liter jugs, and sold it door-to-door as olive oil in working-class sections of Madrid.

First, an eight-year-old named Jaime Vaquero died in pain. Others quickly followed. Prosecutors brought thirty-eight people to trial and asked for sentences totaling 106,000 years. Former Prime Minister Leopoldo Calvo Sotelo had to defend his government against charges that it had covered up details of the calamity. Victims' groups said that up to 836 people had died.

A man named Arcadio Fernández, who suffered toxic effects, kept a plastic bottle of the oil in his closet for ten years afterward. For many, the symptoms included paralyzing headaches and crippled limbs. "If anyone doubts it," Fernández told reporters, "they are welcome to try my oil."

Eventually, thirteen businessmen were convicted of public-health fraud, but they were acquitted of manslaughter. By 1991 the Spanish state had paid one billion dollars in costs and damages. And in 1995 seven former government officials finally went on trial.

Manipulation of oil is one thing. But organized crime has found countless other ways to make huge illegal profits without bothering to adulterate oil. Subsidy fraud is a major industry. The Sicilian Cosa Nostra, like the mainland Camorra, the 'Ndranghetta and the Sacra Corona Unità, make scores of millions of dollars a year in European Union subsidies for trees that do not exist, oil they do not make, and exports that never leave Italy.

Sometimes they smuggle in Tunisian and Turkish oil, not only evading duties but also declaring it as domestic to collect European Union aid meant for struggling farmers. In 1995 Italian authorities seized two vessels off the coast near Bari. They had transferred olive oil of mysterious origin to EU–registered ships and equipped it with phony paperwork.

Greeks and Spaniards are also guilty, and unlike in Italy, neither officials nor reporters say much about it. Corfu probably produces a third as much oil as it declares for subsidy. But in Italy fraud is better organized. Investigators in Brussels estimate that scams in Italy over four years have cost nearly a half billion dollars in agricultural subsidies: wheat, beef, wine, tobacco, milk, and—not last on the list—olive oil. And that is only what the authorities know about.

The European Union's annual budget approaches $100 billion. Half of that goes to agricultural subsidies. Perhaps as much as $5 billion of EU funds are stolen each year; official figures are lower, but no one really knows. Olive subsidies were first devised as a sop to Italy before Greece, Spain, and Portugal joined the Common Market. Now they run close to $2.5 billion a year for payments per tree or kilo of olives, export aid, price supports, stockpile costs, and other assistance.

Of all sectors, olive oil is among the most corrupt. Many crafty independents simply lie and cheat. Farmers declare empty fields and freeway interchanges as olive groves. Millers report oil from producer-clients who died years earlier. Large companies push the legal limits as far as they can. In the 1980s Ferruzzi set up a lobby operation in Brussels that occupied sixty people, taking up two floors of a building.

Not surprisingly, big-time criminals have turned the free-money business into well-organized rackets. Rather than taking the trouble to evade customs agents or enforcement officers, Italian mobs buy them. Or terrorize them. No one, for example, has yet explained why an olive-oil inspector named Antonio Tarsitani turned up full of bullet holes at a freeway service station in June 1993. Similarly, crime families have purchased the loyalty of many legislators and officials who set policies.

Officially, Italy produces 500,000 tons of olive oil in a good year. Fraud investigators believe the real total is below 400,000. For years, olive-oil fraud was one of Europe's best-known open secrets. In 1994, however, a member of the French National Assembly dropped a bomb. In a fat book entitled *Main basse sur l'Europe* (*Hand in the European Till*), François D'Aubert detailed the multibillion-dollar looting of European Union agricultural subsidies in every member state. D'Aubert was head of the anti-Mafia committee of the French Assembly and later a cabinet minister under President Jacques Chirac, and his word carried a lot of weight.

Among other wholesale depredations, D'Aubert noted the case of a large Italian olive-oil merchant who fictitiously exported and imported the same shipment five times, pocketing about $10 million in public funds. At a mill in Sicily, half the suppliers were fictitious; one had been dead for a century.

His main point was not only that fraud existed but also that no one at EU headquarters seemed to want to do much about it. "The Italian code of *omertà* is dwarfed by that in Brussels," he said. The reasons were apparently national sensibilities and a bureaucratic fear of confrontation. This, he said, is not new.

Back in 1985, according to D'Aubert, a fiercely anti-Mafia Sicilian member of the European Parliament pushed through a resolution ordering the European Community to investigate fraud in Sicily. "This investigation was conducted rather strangely," he wrote. It was entrusted to the head of the EC subsidy agency, an Italian who for three years had neglected to send anyone to look into persistent reports of gross violations.

Investigations took four months to get started, and they lasted barely four more months. "A week in Sicily spent in meetings with senior officials of the state and region (which were partly under mafiosi influence) was enough to convince the investigators that controls were perfectly satisfactory in the most exposed areas," D'Aubert wrote. "Only one small concern: the control of subsidies to olive oil; the region of Sicily, which was in part responsible, did not have enough staff to verify the claims in person." According to the investigators, the Mafia was implicated in "only" 10 percent of EC fraud in Sicily, although Sicilians cheated, on average, seven times more than the rest of the EC. No specific cases were cited in the seven-page report, which was read to the European Parliament—and forgotten.

For frustrated Eurocrats, D'Aubert's attack rang true. Follow-up was not an EU tradition. In 1985, for instance, the European Court of Auditors published a sixty-page report excoriating fraud and mismanagement in olive-oil subsidies. The situation got steadily worse. Meantime, the report sank so deeply into the ocean of paper in Brussels that it took me two months to fish it out.

In the midst of Italy's "Clean Hands" corruption trials in the mid-1990s, the authorities in Rome began to crack down. Much of the pressure on the government came from honest olive growers and oil makers who suffered from an undeserved slimy reputation.

Colonel Aldo Coscarella, a dapper cop in a handsome tie and well-cut wool, heads the special unit of Italy's fiscal police, the Guardia di Finanza, which roots out European Union fraud. He coordinates with the Guardia's anti-Mafia investigators and a dozen other law-enforcement units. If a lot of Italians shrug off mob activities as an inevitable fact of life, the Guardia battles away. Morality aside, it's a matter of very big money.

"You've come at a good time," Coscarella told me in his office, near the via Veneto in Rome, in March 1995. "Just yesterday we seized a shipment of 18,000 tons of olive oil that was not what it was supposed to be." That is enough to supply the state of Ohio for most of a year. Labeled as extra virgin, its retail value was 80 billion lire, or $50 million. In fact, only 5 percent of it was extra virgin. The rest was seed oil and refined olive oil of a type suited more to fueling lamps than to cooking.

Thirty executives of ten different companies were arrested. Detectives tracked down the fraudulent oil with help from two men convicted as members of the Camorra, Naples' version of the Cosa Nostra. A year earlier, they had given details about a far-reaching grain scam, naming companies and individuals involved. The Guardia quietly watched all the other dealings of those companies.

"There were certainly mafiosi connections in this, although we don't really know to what extent," he said. "You almost never know exactly who is behind what. We can be sure only when there are confessions, and these people do not like to talk."

Over a two-year period, investigators had seized 54,000 tons of fraudulent oil. A drop in the bucket, authorities said, but things were looking up. After years of laissez-faire and even complicity, Coscarella said, the Italian state is taking the cheating seriously. "All the fraud began to get embarrassing," he said. "Our European associates complained, Spain particularly, and we decided to take firm action."

Coscarella struck me as a good cop, well prepared with his figures, eager to admit failings, and apparently anxious to help coordinators in Brussels cast a wider net. But even if Italy pursues its crooks with new vigor, it will take a lot of cleaning up. Crime syndicates have had centuries of practice in keeping ahead of the law. When the Guardia began sending up airplanes to scout out phony olive groves, someone came up with a handy countermeasure: plastic trees.

Early in 1996 magistrates in Trani uncovered two separate major scandals. One group cut olive oil with seed oil from Tunisia and Turkey, imported via Crete to avoid import duties. Investigators said some of it was purchased—probably unwittingly—by large companies, including Bertolli, Cirio, and Nestlé Italiana, which owns Sasso. In the second case, two French-Tunisian brothers were implicated in the large-scale sale of olive oil mixed with cheap Turkish hazelnut oil. Local newspapers speculated that the Calabrian 'Ndranghetta had a hand in these operations; no one could be sure. For certain, however, both frauds involved accomplices operating in several countries. And police suspected they were on to a major international operation.

"The opening of borders means that law-enforcement agencies can work together, but so can the criminals," Coscarella said. "Whatever we do, they always manage to stay a step ahead."

This was still Italy, where petty larceny is cultural, not criminal. In a country of 55 million people, there are 4 million on disability pensions, most of them arranged by vote-seeking politicians. According to the army's anti-Mafia unit, the BIA, the Mafia is Italy's second-largest conglomerate, after Fiat. And justice is not swift. A year after I saw Coscarella, I called to ask about progress on his case. It was, he said, creeping its way through the courts.

Like other Italians, mafiosi love their olive trees. Apart from the income they bring, they stand for prestige and tradition. That is, legitimacy. In January 1996 *The New York Times* profiled Baroness Teresa Cordopatri dei Capece and her lonely battle with the Mafia near Reggio di Calabria. One morning in 1991 a gunman pumped three bullets into the face of her brother Antonio, who would not sell their thirty-acre olive grove to a local crime family. The killer shot at her, too, but the pistol misfired.

The big trees used to frighten the baroness as a child, when she wandered among their shadows. Still, she said, they were part of the family. The Cordopatris settled in Calabria in the Middle Ages, but their property dwindled over the centuries to a hundred acres, including the trees. Before dying, her father told her and Antonio never to sell the olives. "In these trees," he said, "our past is recorded."

Gradually, the Mafia bought up all the other olive land in Piana Gioia Tauro, near the Aspromonte Mountains. They offered low prices, backed by threats, and often never paid. Only the Cordopatris held out. On Christmas 1990 they made Antonio the offer he refused: "This holiday you are eating sweets; the next one you'll be eating dirt." In July he was dead. His sister testified at the trial, which put the hit man in jail for twenty-five years. One of the men who hired him got a life sentence. Now, at sixty-three, with a full-time police guard, Baroness Cordopatri campaigns for official action against the Calabrian mob.

She has yet to pay $80,000 in inheritance tax, because the Mafia has scared off her olive harvesters. "Why should Italians pay taxes when livelihood and land are off limits to them?" she asked. "It is not enough to put a few criminals in jail. People in Calabria must believe they are safe from the Mafia . . . If I cannot work my land, then it means the Mafia is still winning. It should be the mafiosi who are under guard, not me."

The baroness appeared on a series of television interviews in Rome, but came home pessimistic. "Things in Italy do not change," she said. "Our problem is thirty years old, and no one ever lifted a finger to help."

As 1996 began, sentiment across Italy was turning against magistrates who applied tough measures to root out corruption. Many people were charged, but few went to trial, and fewer still went to jail. Vittorio Feltri, editor of the conservative daily *Il Giornale*, cracked to an American reporter: "If you cut off the right hand of everyone who has paid the Guardia di Finanza, you'd have a country of left-handed people."

The Clean Hands campaign was a sobering experience. People expected politicians to be implicated *en masse*, but even the cynics were surprised. Former Prime Minister Bettino Craxi was sentenced to eight years in prison. One former health minister was jailed for taking a regular cut of money allotted for medicine, hospitals, and AIDS victims. One of his senior officials was found with $100 million in cash, fourteen Swiss bank accounts, a hundred gold ingots, and assorted houses. He had, he admitted, "bent the rules a little."

And then the police arrested Giulio Andreotti, seven times a prime minister, a member of thirty governments, and a man who played a major part in shaping postwar Italy. He was indicted for collusion with the Mafia, and he is accused of sanctioning two murders to cover up corruption. The trial could take a long time.

In a burst of righteous fury, Italians elected Silvo Berlusconi as prime minister, a media magnate and among the country's richest men, whose coalition, Forza Italia, was finally to put things right. That government had lasted seven months when Berlusconi was charged with corruption.

Who would stop at olive oil?

The umbrella organization above Coscarella's operation is known by its French acronym, Uclaf (Unité de coordination pour la lutte antifraude), the European Union's anti-fraud agency. Under EU legislation, a single agricultural policy applies to all member states. Guidelines cross borders, with no exceptions. Governments enforce European law in their own territories. And Uclaf is supposed to make sure that each works with equal vigor.

Uclaf is headquartered in one of those building clusters at the edge of Brussels which can induce narcolepsy in an insomniac. The real world penetrates by modem, fax, métro, and mail, but its heartbeats and bodily fluids seldom get past the gray-faced guards. I arrived with an appointment to see the director, a Dane. But a television crew from Denmark had shown up, and I was blotted from the agenda. Instead, I saw Siegfried Reinke, his German chief adviser, and a pair of French investigators who specialized in olive-oil fraud.

Reinke is the sort of German official beloved by Hollywood film casters: commanding presence, mysterious scar, metal-rimmed glasses, and sardonic smile. A well-thumbed copy of D'Aubert's damning book lay in his in-basket. "He considers it one-third fact, one-third exaggeration, and one-third fantasy," an aide remarked in an aside. Reinke confirmed its main lines. People cheat outrageously on subsidies. Some are crooked individuals. Others are mafiosi. All are hard to catch.

Uclaf's new computers, IRENE and SCENT, make the job easier. They spot names which recur often and track suspect shipments across member-state lines. Patterns emerge, and history comes back to haunt. Tough little teams work with national authorities to improve law enforcement. Uclaf, for instance, has people working with Coscarella's squad in Rome. Brussels sniffs out leads and passes them on to local cops.

"But," Reinke acknowledged, "it is a very big problem." Uclaf can help train and support national anti-fraud squads and coordinate them. But its investigators cannot make searches, seizures, or arrests. "We know that a lot gets by us." There is a particular problem here. If a member country reports fraud, it assumes the responsibility of tracking down the culprit and recovering the money. National agencies are short of staff and underfunded.

Uclaf's 1994 annual report notes that its investigators handled a total of 1,597 fraud cases, involving $320 million. Of that, 507 cases were closed, but only $16.6 million was recovered, less than 10 percent of the 1995 appropriation for anti-fraud activities. In olive-oil fraud, only 17 of 171 cases were closed. Of the $47 million stolen, $2 million came back. These were the worst results among the twenty-seven sectors listed.

Another table showed what percentage of known defrauded EU funds were recovered within member states. This is a touchy issue, since Brussels officials say privately that countries are less apt to protect community funds than their own. Denmark did well, getting back 41 percent. Italy recovered 1 percent. As for Greece, with much less reported stolen, the figure was zero.

Along with a stack of reports, Reinke handed me a cutting from the London *Sunday Times* headlined: EUROPEAN OIL SCAM: WHO WANTS TO VOTE FOR THIS? The British, cool enough to an overbearing common market, were not thrilled at southern Europeans stealing so much of their taxes. D'Aubert had made the point after his book appeared: "European money is not considered public money. It's money without nationality, which means it is anyone's to take."

Anita Gradin, a Swedish stateswoman who was the EU commissioner responsible for fighting fraud, held a spirited news conference early in 1995. "The misuse of [EU] funds was of the highest importance," she said. She increased the Uclaf staff to 130 people. But when I saw her later that year, she seemed almost resigned to an impossible task. Like Reinke, she acknowledged that investigators could see only the small tip of a large iceberg.

The problem goes beyond Uclaf. European Union officials, in effect, answer to fifteen mothers-in-law. The bureaucracy is plodding, and it works in nine languages. They can plead with member states, or maneuver, chide, and bargain with favors. *In extremis*, they can impose sanctions. Late in 1994 the commission threatened Greece with a $700 million fine if it did not stop cotton farmers from falsifying their production claims. But that invites confrontation, which Brussels abhors; the Greeks accused northern states of discrimination and harassment.

Corruption among national authorities, of course, makes enforcement difficult. At Uclaf I mentioned an instance I had heard about in Greece. A Kalamata oil merchant had received a call from a man in Athens with high-ranking official connections who wanted several truckloads a week sent to Bulgaria. The merchant said he could not handle that much olive oil. You don't have to, the caller replied. Just send the paperwork. His friends in Greek customs would see the trucks pass by. With stamped documents, he could collect EU export subsidies of nearly a dollar a kilo. An investigator asked for names. I did not give them, and he didn't press the issue. It was only one more untouchable mystery.

The worst problem for Uclaf agents who specialize in olive oil is mislabeling and adulteration, and they can only guess at the magnitude of the problem. In preparing its olive-oil report, the Court of Auditors sent someone to examine stockpiled Italian oil. Of the twenty-three samples taken, twenty-two were of a lower quality than their labels declared. This was surprising, the auditors said, since Italy had just reformed its classification and labeling controls. The Uclaf report for 1993 also noted a sampling of olive oils stockpiled in Italy. The amounts tallied with official numbers, but 93 percent of the oil was of a lower quality than declared. A third of it was of such an inferior grade that it did not qualify for price support.

"We can't check it all," one Uclaf investigator admitted. "You can clear one batch, but the next one might be bad. Even accidentally, you can have bad oil. Chemically it is very hard to tell what is extra virgin and what is not. When people try to deceive you, it is all the harder."

Recent chemical refinements have made it easier for adulterators to cover their tracks. If the characteristic sterols are removed, rape oil is extremely difficult to detect in olive oil.

Finally, I called Joe Carey, an Englishman who resigned from the Court of Auditors in 1993 because of its inability to curb fraud. He continues to do consulting work with the EU in Brussels. The Italians cheat enormously, he said, but the Greeks may be even worse. In Greece the problem is not so much organized crime as larcenous opportunists. While Italy has a land register which specifies who owns which trees, Greece does not. People claim subsidies two or three times for the same olive grove.

"Nobody has even scratched the surface of the biggest problem," Carey said, "which is the lack of motivation for member states to enforce the rules. The European Union is built on the fiction that there is a difference between your money and my money. This is an enormous hurdle that no amount of Uclaf activity is going to overcome."

If there was ever a case of *caveat emptor* this was it. Just as Armando Scaramucci warned at Forci, when you buy olive oil you have to know with whom you're dealing. Wrongly labeled oil can ruin pleasure, and possibly stomach linings. Undiscovered large-scale fraud can drive honest producers out of business.

There was more to learn, but other subjects pressed. That was probably just as well. Puzo's *Godfather* still influenced modern wise guys. I was not anxious to find a severed section of Emiliano's beloved old trunk lying on my pillow.

Marseille Soap

Pavé de Lotte aux Olives de Nyons
et Son Confit de Tomates

At his Le Relais de Grignan near Nyons, Hubert Batin has found a delicious way to add tart black olives to delicate white monkfish. The Nyons
tanche is perfect, but any good, small-sized, salt-cured, but unflavored black
olive will work.

1 7-ounce piece of monkfish per serving (other ingredients to proportion) Several pitted small black olives	Extra-virgin olive oil Several ripe tomatoes Garlic cloves (one per tomato) Fresh basil to taste Salt and pepper to taste

Make a ¾-inch incision across each piece of fish and stuff it with
olives. Tie the fish with a string, *tournedo*-fashion. Cook 10–12
minutes in a skillet with some oil (Batin recommends Nyons, of
course), then set aside, keeping it warm and covered. Blanch, peel,
and deseed the tomatoes. Heat more oil in a saucepan and cook
the tomatoes until they are reduced to a purée. Add a clove of
chopped garlic and a pinch of fresh basil per tomato. Add salt
and pepper to taste. Cover and cook 20–25 minutes at low heat.

Garnish the plate with the tomato *confit*. Place the fish in
the center, decorated with basil leaves and a few pitted black
olives. At the last moment, drizzle olive oil atop the fish.

"The olive orchard is like a library, where
one goes to forget life or to understand it better.
In certain villages . . . where there is no other
distraction than solitude, the men, on Sunday morning,
go to the olives just as the women go to mass."
                                        —Jean Giono

8

Running of the Olives

Meanwhile, back at the orchard, Emiliano and the others were doing well.
I had pruned, plowed, poisoned, and prayed. New shoots poked upward
from amputated limbs. Tender green branches promised a decent crop for
the next season. Sheep manure was heaped next to a woodshed bulging
with garden tools. Wild Olives was taking on a tamer look.

And I was getting obsessive. On long walks I stopped every few yards
to worry over someone's ailing or misshapen tree. A visiting friend, who
had lived among the Sioux and the Navajo, gave me a name: Talks About
Olives.

But my own visits were brief and infrequent. I had learned another
of those Provençal homilies: He who has only olives will always be poor.
Mostly I was at work in the real world. One January I prepared for an
assignment to Havana, where Hemingway spent his later years musing
about the running of the bulls in Pamplona. By chance, I came upon an

item in *Le Monde*: Nyons, a small medieval city near the Rhône, was about to hold the running of the olives. I just had time to make it.

Nyons, in the Drôme just above Provence, was having its annual Fête de l'Olive, which celebrates the new oil. Along with tastings and speeches, the fair featured a five-mile foot race through the streets and up among the thick stands of gnarled trees that covered its flanking hills. Pamplona's annual dash produced more injuries but hardly less enthusiasm. Although no one was gored by an olive tree that year, several people were hurt in a highway pile-up as they hurried into town for the festivities.

All Mediterranean cultures celebrate their olive harvest, but no one takes it to heart like the French. The largest mill in France, Jean-Marie Cornille's seventeenth-century press at Maussane-les-Alpilles, near Les Baux, crushes one thousand tons of olives a season. That's no more than a mere weekend's work for Francisco Gutiérrez Lopez's plant near Jaén. But no matter, pilgrims come from halfway around the world to make appreciative noises.

In France, olives and their oil, taken to extremes, fall into the category of *grand luxe*. You can speak of them with intimidating finality, your nose angled upward. Mainly, they are something to be put in the mouth. Olives, as any real Frenchman can attest, belong to France. And the small black Nyons tanche is France's only olive protected by law with an *appellation d'origine contrôlée*.

Nyons is not on the way to anywhere, and that is part of its charm. It is just south of the Olive Line, which starts in the small garden outside the McDonald's at an autoroute stop near Valence. Rarely do its patrons notice the three healthy trees growing outside the window.

To get to Nyons, you continue down the A6 and turn east at Montélimar, about an hour south of Lyon. The narrow road changes directions and route numbers, occasionally ditching inattentive drivers who focus on the scenery. This is rich, heartland France. The old castles look Teutonically solid but southern fanciful. Nyons itself, on the spirited Aigues River spanned by a Roman bridge, winds upward along narrow streets overhung with balconies.

I reached town at dark and stopped to investigate a blaze of lights at the Nyons wine and olive cooperative. Inside, Freddie Tondeur held court among a crowd of growers and dignitaries. He wore a green velvet cutaway cloak, which gave him the air of some Jurassic lizard. A matching Alpine hat sprouted an olive sprig. Tondeur, a local writer, was grand master of the Confrérie des Chevaliers de l'Olivier.

Heaps of black tanches, flavored delicately with herbs, quickly disappeared from the tables nearby. Bottles of wine emptied at an equal pace. Suddenly a hush fell over the room. In excited little knots, the crowd moved into an adjoining vaulted chamber and mobbed a long, white-clothed table. Clay bowls were piled high with fat cloves of garlic. Baskets overflowed with baguette slices. Plastic spoons lay in bunches. And in clear flasks lined up next to shallow dishes, the year's crop of Nyons oil was ready for tasting. Elbowing in with the rest of them, I took a piece of bread and then rubbed, poured, sniffed, scarfed, and smiled. The tanche, grown in cooler, higher places, gives lighter oil than other French varieties. When fresh from the mill, it is crisp, with a full blast of fruity bouquet. By the year's end, it is heavier and flatter, like most aging oils. When a new crop is pressed, Nyons is ready.

The scene was repeated the next day for the masses in the Place des Arcades, an old Nyons square bordered by elaborate archways. People browsed among booths, buying olive-motif potholders, carved olive wood, olive-patterned shirts, and young trees. They watched a sizable pack, from teenagers to old men, take off on the race. On a large stage, folk groups sang and danced homage to their beloved tanches. The crowd applauded when the runners came back and sat patiently through speeches extolling Nyons as the fount of all good oil. Then, with great ceremony, recently elevated chevaliers received the green cloak of rank.

Shortly after eleven o'clock, Tondeur invited the crowd to taste the new oil at tables by the edge of the square. They responded with a cheer and a stampede. I sneaked around from the back to take pictures and found myself in a privileged spot. Amid the tumult, I had a clear two feet of table frontage, with a bread basket, a garlic supply, and a bowl of oil.

All around me, the horde attacked. Some rubbed the garlic carefully into the bread and neatly spooned on the oil. Others ignored the garlic and mashed their bread into the open dish. After a half dozen samples, I moved back to snap a few frames. When I went back for more, a roiling cluster of newcomers had claimed my territory.

But then there was lunch, the traditional *aïoli* feast for the chevaliers and their guests. Each person was served a heaping plate of salted cod, potatoes, and boiled vegetables, along with a mound of *aïoli*, a thick mayonnaise of olive oil, egg yolk, and crushed garlic.

The business at lunch was the ninety-ninth session of the Confrérie, during which thirteen new members were inducted. Each was knighted with an olive branch and took the oath: "I promise to defend the Olive Tree and all the true riches it offers, material and spiritual; to practice the virtues it represents; to aid with all of my means in maintaining and promoting its culture; to work for the Olive Tree, nourishment and light, for the Olive Tree, symbol of abundance, of wisdom and of peace, symbol of life."

Two professors of archaeology, old friends who had explored the olive's early role in ancient Crete, were thus knighted. An Italian mayor, an olive grower, a tourism executive, some businessmen, a marathon runner, and a journalist followed them. Among the sober suits and serious ties was a young inductee in ragged shirttails, sorely in need of a haircut. He turned out to be Jean Giletti, a Nyons cardiologist who had studied why French oil-eating country folk live longer, in spite of all the greasy lamb, pork salami, and egg yolk in their diets.

Tondeur delivered a stirring tribute to the tanche, to its unique AOC status, and to Nyonnais growers. "For a long time now, twenty centuries at least, we have known that olives belong to our patrimony and that by a happy conjunction of sun, land, and local know-how, our fruits are the best in the world." The olive, concluded the grand master, symbolizes peace at a time when the world is suffering from hatred, intolerance, and war. "When will men put an olive branch in their hearts?"

That morning, the running of the olives turned out to be a tough uphill race. Halfway through, I gave it up to stop and admire a tight little grove of forty-eight old trees. Tanches are normally shaped on a single trunk, which corkscrews with age. After the first century, they are venerable sculptures of twisted burls and bulges. I did not notice the old man behind me, who demanded to know who I was and what I wanted.

Georges Vallon, husky and square-shouldered at seventy, eyed me carefully through thick glasses. With explanations made, he warmed to a friendly glow. Vallon was a retired farmer whose trees were his life, and he was thrilled that someone was curious about them.

He had packed good dirt with manure high up around the base of each tree. That, he said, keeps the root mass well fed, humid, and protected from the cold. He showed me the male pollinating tree—pointy leaves, pink-tinted olives—with all the pride of a poultryman boasting of a prize rooster. "You have to let the horizontal branches grow out," he said. "Vertical branches don't produce." Some vertical branches do grow olives, but I did not argue. Those broad, full crowns were no accident.

Vallon picked each olive personally. "It takes a long time," he said, "but I'm not paid by the hour." He grimaced when I asked about the new mechanical pickers. "*Ah, oui,*" he said. *"Les branleurs."* The word, meaning shakers or vibrators, is a vulgar term for male self-abuse. By then I'd heard a lot of growers speak with enthusiasm about the vibrators. Hard shaking invigorated the roots, breaking some old underground runners to make room for new ones, just like pruning the crown.

He talked on about the isolated corner of rural France he had inhabited for forty years and delved into the comparative virtues of various animal droppings. He seemed as permanently rooted as his trees. Then, for no particular reason, Vallon said, "On Saturday I'm going to Cuba."

Vallon loved to travel. He had been everywhere, from China to Chile, in search of beaches and mountains and the world's wonders. He always kept an eye out for olive trees. "Tunisia, now there's an olive country," he said. "Beautiful trees, big ones, whole fields of them. Tunisians pick the olives with a goat's hoof." A goat's hoof? "Yes, of course," he said, smiling at my ignorance. "They take the foot and rake it along the branches. It works marvels."

At night in Nyons, I was directed to a small restaurant called Le Petit Caveau. It was nothing special, just a run-of-the-mill wonderful French provincial restaurant. Sitting before the sparkling glass and silver, unfolding a crisp, colorful napkin, and discussing the fine points of dinner with a teenaged waiter in black tie and tuxedo, I reflected on the olive world.

In Baena, the fanciest restaurant we could find was rich in charm and history, a converted Guardia Civil headquarters built centuries ago. The headwaiter was so friendly and helpful that we had to smuggle the awful food from the table in an empty cigar box to keep from hurting his feelings.

Juanito's in Baeza and the Núñez de Prado feast were the exceptions that made the rule.

In Italy the meals were invariably tasty and, just as invariably, set down on an unadorned table in a manner suited to no-nonsense nourishment. The cuisine ranged from good to very, very good, but it was short of surprises.

In Nyons, as elsewhere in Olive France, the chef had his self-respect on the line with every tip of the cruet. Monsieur Cormont's opening aria was duck carpaccio marinated in Nyons oil and brushed with *tapenade*. That was followed by a *marmite de la mer rotie*, succulent bits of seafood with celery and leeks lightly fried in olive oil. The cheese course was a sweet-sharp local chevre doused in tanche oil.

The hand-lettered menu suggested that diners might wish to leave wine choices to Muriel Cormont, sommelier-conseil, diplômée de l'Université des Vins de Suze la Rousse. It was sound advice. With the duck, Madame Cormont brought a glass of lively Sablet Domaine des Gouberts Mr. Cartier 1989. A different Sablet came with the seafood.

I expressed such appreciation, and feigned such familiarity, that she scurried off for a bottle she kept in some dark back corner. "I'll tell you later what it is," she said, with a wide smile, playing me the compliment of pretending that I was capable of identifying some obscure Coteaux du Languedoc.

Cormont, in fact, is no particular devotee of olive oil. He learned to cook with butter in the north, but when he moved to Nyons, he knew what his customers wanted. He experimented with the subtleties of adding, say, grapefruit juice to olive oil in new dishes. Soon enough, he made cooking with the tanche his new art form.

French oils, for many connoisseurs, are the best in the world. Charles Aznavour remembers how for decades he brought Frank Sinatra an annual supply of peppery but sweet oil from the Golden Triangle, bordered by Carpentras, Maussane, and Aix. If he ran late, Sinatra was on the phone howling.

Picked later than Tuscan varieties, French olives are richer in oil content, and the oleuropein has softened naturally. Andalusian oils, most typically, taste strongly of olives. Oils in France range from spicy to buttery, and yet each is good for almost any purpose, from delicate salads to frying fish.

Nature and milling techniques account for the basic characteristics. But French chefs will argue that their own favorite oil is best for the same reason they excel at anything that touches *haute cuisine*. In the end, where taste is concerned, passion matters nearly as much as palate.

After the 1956 freeze, the Bouches-du-Rhône replaced the Var as prime French olive country. This is the chic section of Provence, near Saint-Rémy, dotted with stately homes of pink-hued stone and country lanes shaded by towering elms. Olive mills are revered as places of worship.

At the Maussane-les-Alpilles mill, Suzy Ceysson dresses up for work as though she was marketing manager for Moët & Chandon. She grew up in the groves, tending trees her grandfather fretted over, but only the little gold olive-and-leaf cluster on her lapel gives her away. I went to Maussane because of a recent *Washington Post* article about "Jean-Marie Cornille's mill." When I asked to see Monsieur Cornille, the mill hands looked at me as if I was nuts. Cornille, one explained, had been dead for four years. Local olive people loved the old miller so much that they left his name on their cooperative press. The new director was Suzy.

She wasted no time with polemics. Her oil was good because the region produced three exceptional varieties of olives, and her mill pressed them quickly and cleanly. For proof, all anyone needs is a teaspoon or a crust of bread. Roger Vergé of Le Moulin de Mougins, near Cannes, not one to trifle with his three-star reputation, sells Maussane at the risk of offending producers closer to home.

Maussane is one of the few French oils offered by Dean & Deluca or other American specialty shops. This is partly because of its high production, for France, and aggressive marketing. Each month, Ceysson sends 3,600 bottles to Tokyo and another big order to Osaka. The mill was featured on Kansai-TV, and now buses full of Japanese regularly tour the place.

The Maussane process is a grander version of the Gervasonis' in Aups. Oil streams down the tower of mats. Water is spun off by a centrifuge, but the oil is not filtered. If you visit within a few months of pressing, you have a choice. The previous year's oil is clear and mellow; lingering sediment has settled to the bottom, and most of the sharpness is gone. But the new oil is different. It is fresh and fruity, with a peppery bite. Tiny bits of olive make it cloudy. Unsettled from spinning, it is troubled oil. Or, for the French, *huile trouble.*

"Around here, olives are a way of life," Suzy said. Villages in the area have their yearly festivals, smaller than Nyons' fair but no less enthusiastic. "If you grow up with the culture, it shapes you in a certain way."

But one of France's finest oils, Saint-Côme, flouts all prejudices. Nicole Allione, the miller, was a postal worker in the mountains before she married her Provençal husband. Her olives are picked in November, green, not black. They cover a wide spread: aglandeau, an Haut Var variety, are mixed with salonenque and verdal, the Alpilles stalwarts. While everyone else presses in *scourtins*, she uses one of the Alfa Laval continuous systems that seemed so out of place at Frescobaldi.

Madame Allione's oil was awarded a gold medal by Max Doleatto and other judges at the 1995 Paris agricultural fair. It is full-fruited, with a greenish snap underlying the usual French sweetness. "There is no mystery about it," she says. "If you use fresh, healthy olives, your oil is good. If the olives are bruised or wormy, it is bad. We refuse a lot of olives from local growers and make a lot of enemies. Too bad, that's how it is."

The squeaky-clean mechanical mill is designed to be visited, complete with a crushed-gravel parking lot. A vast tourist shop offers ceramic kitsch and overpriced sacks of *herbes de Provence* tied up in colored ribbons. I loved the oil, but I missed the passion. Madame Allione had mastered a business, and she was available if anyone wished to make a purchase.

The French penchant to colonize anything related to the taste bud is often done with such flair and conviction that one is tempted to accept it on faith alone. Sometimes that is justified. Sometimes not.

On the rue de Rivoli in Paris, A l'Olivier is fitted out with shelves of handsome wood and markings stamped in gold. The overwhelming impression is of a temple to olive oil, served by a devoted order of *religieuses* steeped in its catechism. This, I found on closer inspection, was not necessarily the case. One afternoon the woman in charge fended off questions about her oil with a dismissive smirk, and then sneered at a customer, a pleasant Frenchwoman who asked for Tuscan oil. The other extreme is far more common. Most French olive people see themselves as missionaries, patient and kindly, eager to enlighten benighted butter people.

In fact, A l'Olivier's oils are good: light and golden. They come from around Nice, where moist Mediterranean air among groves of cailletiers produce a mild oil. They do not do much for steamed vegetables, but are excellent for everyday use and perfect for seafood.

A distinctive oil of a different sort is made by Nicolas Alziari. It comes in a striking can: olives and leaves painted on a background of rich gold stars and against the deep blue of a Provence sky. It has a nutty aroma and a fresh fruit taste. Alziari's shop by the old Nice port offers an array of table olives, but I like the mill.

Nicolas Alziari's son, Ludovic, died in 1996 at the age of eighty-seven, pressing olives to the last. He had sold the business in 1990, but he lived upstairs and came down to work every day out of sheer habit. "He was the pope of olives, and this is the mother church," his acolyte, Patrick Kioutsoukis, told me. "Well," the young miller added, with a glance at the large table in back splashed with red wine and oil, "at least it's where all the old olive people party after the harvest."

The old Roman process—*à la génoise*, in French—was something I had never seen. Stones mash the olives in deep vats made of polished limestone. Water is piped in, and the oil floats to the top. The olives' owners rake off the residue, and Kioutsoukis collects the murky oil in a metal can. After decantation in a separate stone tank, the oil is ready. That is for the growers' private stock. When the water is removed, the remaining paste is pressed in coconut-fiber *scourtins* on a traditional tower, and it is decanted in a 600-year-old limestone tank.

Kioutsoukis, Greek-born, was a waiter until Alziari taught him the business twenty years ago. He explained his work as he deftly lifted oil from the tank with a specially shaped scoop, like a miner panning gold. He must have read my mind. "When the rays of afternoon light come in that window and strike the oil," he said, "you would swear it was gold."

The new owner is Jean-Marie Draut, a vintner who decided to make oil five years ago. "It was the logical move, a full circle," he said. "Alziari used to say that what killed olive oil in France was Verdun, the 1914–18 war, when so many young men died that there was no one left to work the trees. And I came here from Verdun."

Draut's mill makes only forty to fifty tons a year, and distributors clamor for it. "The way prices are these days, it is getting to be like selling drugs." Once, a high-energy New Yorker burst into the office and started to order. "He wanted a pallet of this and a pallet of that," Draut said. "I asked if he had tasted the oil, and he said no. I told him to forget it, we're not selling Coca-Cola."

Naturally enough, Draut is convinced that his oil is the world's best. It lacks bite, and he does not miss it. "Olive oil should be part of a dish," he said. "It should never overpower the taste of what it accompanies."

For me, the best French oil is not from Nyons or Nice but rather the hand-separated, down-home product of our own rocky hills in the Haut Var. There is, for instance, Fabrice Godet's Moulin de l'Horloge in Draguignan. It has nothing unusual, like Alziari's *génoise* vats, and it is another world from mills in the posher parts of Provence. Clean where it counts, it is otherwise a funky jumble of tattered ledgers, oil barrels, and certificates of merit affixed to the rock walls.

Fabrice is humble about his work. Once, I dropped in and mentioned that I'd been to Baena to see the Núñez de Prado brothers. His eyes widened. It was as though I'd told a country imam that I had just returned from Mecca. "Ah, Núñez . . ." he said, giving the name a lilting French twist. "If there is a high priest of olives," he said, "it is Andrés Núñez de Prado." But when he presses his own olives, Fabrice's oil is in the same league. And he makes his own *flor de aceite*, the old way. He scoops up the puddles of raw oil that seep out after mashing.

Young trees are supposed to yield better olives. But our old trees have seen everything, and every year they give a little more. Six months after pressing, our neighborhood oil is still murky with flavorful fragments that cloud it to a golden green. We reach for the Tuscan when in need of a sharp nuance. Special oils from other places all have their purpose. But for overall taste, and maybe just for the idea of it, we love *huile trouble*.

———

In June our mountainside is at its best. The brooms are in flower: great, dense sprays of brilliant yellow, which perfume the valley in a heady scent. Cherries and mulberries are ripe. Fields are scarlet with delicate poppies and fringed with wildflowers of a dozen hues. And the olives, fresh with new growth, are too full of blossoms to be sprayed for bugs. It is a time for pure enjoyment. As it turns out, it is also a time for birthdays. Three of the Romanas were born in June, and so was I. So, one year, we began a new ancient tradition: the annual Gemini Olive Feast. Our first was at Lucie and Roger's old stone house, lost in what was a medieval wheatfield on the other side of Ampus.

As in Nyons and everywhere else that French oil is fêted by the people who make it, the centerpiece was *aïoli*. Lucie, Annie, and Marie whipped up a bucketful at the Martins' old stone house on the other side of Ampus. Their recipe is dead simple. For each four or five people, take two cloves of garlic, one egg yolk, a little salt, and a half liter of oil. Then beat it all together until your wrist throbs from the pain.

Several dozen of us sat down along a long trestle table set with an assortment of plates and silverware and studded with bottles of a rich Provence red. Bowls and platters appeared in a steady stream: hard-boiled eggs, potatoes, steamed carrots, fat Burgundy snails, and their smaller, tastier Provençal cousins, salt cod. Each forkful was dipped in a pale-yellow dollop of *aïoli*.

Lunch went on for hours. Jeannot ate, hurried off to the town hall to help supervise the mayoral election, and came back to find the food still coming. Old man Romana sat quietly in a corner chair, happily chain-smoking his brown hand-rolled cigarettes. Up and down the table, the topic was olives.

Fernand Fabre, a friend of the family, had been assaulted by olives, and he barely survived. While operating the Cotignac mill, he was slammed against a stone wall when a badly stacked tower of *scourtins* squirted sideways under pressure. Another family friend preferred pigs to motor tillers; he was a truffle hunter, whose intense interest in olives was merely culinary.

Roger Martin explained how olive trees were pickier about where they grew than their reputation suggested. His trees, like his father-in-law's, his brother-in-law's, his son's, and mine, had southern exposure, shielded somewhat from the fiercer mistrals. They were a shade lower in altitude. At his home on the north side, not even Roger could get a tree to thrive. Nor could he do much farming. His driveway passed by an ancient oak, registered as a national monument, and its spreading limbs were too low to allow a truck to pass.

Mostly, I listened to René Pellet, an oliveman with a white Solzhenitsyn beard, who had retired into his grove overlooking an ugly shopping mall outside Draguignan. He was born among the olives and, as a kid, spent each harvest camped beneath the trees with his parents, sleeping in a banana crate. A single word—like bug, or graft, or Greeks, or sheep droppings—launched him on a monologue that not even snails in *aïoli* could cut short. He was, above all, a master pruner in the French tradition.

The Spanish, mostly, go for volume, letting trees grow high to fill with olives that must be harvested by sticks. The Italians are edging toward convenience, shaping their trees short and bushy for mechanical shakers. The French lean toward their classic manner. A tree is formed in its first years, cut back until a single trunk thickens. Three *charpentes*, main branches, support the crown. These are continuously topped, so the tree grows solid but seldom higher than fifteen feet. Olives are the purpose. But the tree is also a thing of beauty.

"Sometimes when I finish, I see all the branches lying on the ground, and it pains me to think of the suffering I've visited upon a tree that has gotten to be my friend," René said. "But then I look at the tree, spare, beautiful, ready for something new. I feel like I've just coiffed the hair of a movie star."

By the time mounds of cherries replaced the *aïoli* fixings, René was feeling his olives. With all of us in rapt attention, he read some of his poetic prose, first in Provençal patois and then in French. He evoked his earliest days, when in one season his family picked enough olives for 1,200 liters of oil, close to six tons. He described the subtleties of the trees, depending on the time and weather. Sometimes the scintillating leaves were schools of little fish, silvery in clear water. In the setting sun, the spreading branches faded into pastels, and then dramatic dark silhouettes, and vanished until morning.

And Pellet expressed bitterness at seeing so many of those treasured old trees, which demand so little to thrive, die of simple neglect.

"Today," he concluded, "when I see people dressed as cabinet ministers, not to say lords, plant a little olive tree in a luxurious garden, I like to remember the old times, closing my eyes to see again those peasants, in rough homespun, tend with wisdom and love those great trees dripping with olives."

Late in the afternoon, the wine bottles were empty, knocked over like bowling pins on the battle-scarred tablecloth. Annie rummaged in a bag and found a decades-old cassette of her father playing his accordion. Everyone danced, whirling around under the grapevines and apricot trees. By then, old man Romana had moved to the head of the table with Françoise. Out in the clean air, stuffed to bursting with oil from his own trees, while his progeny and pals two-stepped to airs he remembered from a distant past, he wore the contented look of a very happy man.

That same summer, Draguignan had its forty-second annual Foire de l'Olive. After Nyons, it was not much. It opened with a solemn march from the subprefecture, including teenagers wearing Provençal costumes they normally would not be caught dead in. Growers sold olives, oils, and trees. One man demonstrated a portable press. Another showed how to graft buds. It was mostly an all-purpose fair for local merchants and traveling hucksters. But there was the olive pit–spitting contest.

In the spring of 1994, a young man spit a pit 11.4 meters in the parking lot of an Arles supermarket and entered the *Guinness Book of World Records*. In the summer of 1995, Draguignan set out to top that mark. Each morning passersby tried their luck, three pits per spitter. By Friday, six days into the fair, a hundred hopefuls had logged in. The results were pretty anemic. Then a gendarme sailed his pit 11.3 meters, only ten centimeters short of the world record. Saturday, on duty elsewhere, he did not return. Tension was high for the Sunday finals.

Only ten of the thirty eligible contestants showed up, for some reason, and the cop was not among them. But Gilbert Fitoussi, twenty-one, who had come from Marseille to sell bed linen, sauntered up to the line. He was coolness personified: pointy cowboy boots, black Via Veneto trousers, a silkish shirt unbuttoned to the navel, exposing a gold Star of David. Gleaming black hair was slicked back over handsome Gypsy-like features.

Gilbert chewed and then calmly drew in several deep breaths. He rocketed to the launch line, cheeks billowed, and spit. The pit bounced at 11.70, thirty centimeters past the record. I watched a radio announcer hurry to him with a microphone. He wore an orange lion's mane, charcoal whiskers, turquoise pantaloons and cape, and gold lamé boots. This was heading a little beyond my subject. It was time to look for those Tunisian goat's feet.

Olive Pickers' Goat Horns

Tunisian Fish Fillets *with* Harissa *and Black Olives*

Tunisians like it hot, and one of Paula Wolfert's favorite recipes is fish and olives in fiery red *harissa* pepper paste.

1½ pounds firm white fish fillets	1 cup tomato sauce
Salt and freshly ground pepper to taste	½ teaspoon *harissa*
Flour	1 bay leaf
¼ cup olive oil, for frying	1 cup black pitted Tunisian or Greek "colossal" olives
½ cup chopped onion	Juice of ½ lemon, or more to taste
2 garlic cloves, minced	Chopped parsley

Season the fish with the salt and pepper; dust with the flour. Heat the oil in a large skillet and fry the fish until golden brown on both sides. Transfer the fish to a warm dish. Pour off all but 2 tablespoons of the oil. Add the onion and garlic to the skillet and cook, covered, 2–3 minutes. Add the tomato sauce, ½ cup of water, the *harissa*, and the bay leaf. Cook 10 minutes. Add the olives and fish fillets and continue cooking, uncovered, until the fish is tender and the sauce is thick. Add lemon juice to taste. Serve with a sprinkling of chopped parsley.

Serves 4–6.

Adapted from *Mediterranean Cooking*

—
9
—

Socialized Olives

Georges Vallon, the Nyons oliveman–traveler, was wrong about goat's
hooves. It was horns. At some point during the last three thousand years
in which they have been growing olives, Tunisians devised a clever way
to remove them from trees. They saw off the last three inches from the
pointed horns of a young goat. Then, slipping the curved and hollow tips
over three fingers of the picking hand, they claw the branches at lightning
speed. Olives tumble onto a tarp below. Neither fragile bark nor skin suffers
in the process.

No one else uses this method. In fact, not many European growers
even realize that Tunisia turns out a delicate golden oil that ranks with
the best. Or that at the edge of the Sahara desert, where rain often neglects
to fall, a single well-tended giant chemlali might produce 800 kilos of
olives, nearly a full American ton. Its close relative, the French cailletier,
seldom approaches fifty kilos across the Mediterranean.

Even on the dry plains of Andalusia, European trees are normally combed by gentle spring winds and watered when necessary from underground streams. If neighborhood flocks don't provide enough manure, subsidized farmers can afford a few hundredweights of nitrogen-rich fertilizer. The adaptable *Olea europaea* performs near-miracles in Tunisia. Great trees, such as the zalmati, chaal, chetoui, and meski, produce richly flavored olives with little water. And yet Tunisia's reputation remains as Dino Fontana described it: a mass producer of crude oil.

"Yes, this can get exasperating," admitted Bechir Ben Maad, whose finest extra-virgin oil was hosed into tankers and shipped off to be blended and sold under fancy Italian labels. Buyers laugh when he pleads for some tiny mention, anywhere, of Tunisia. But he did not sound exasperated. He was used to living on the wrong side of the Mediterranean, in a country that treasured art and literature when most Europeans lived in caves.

I had spoken to Bechir at Oleum, in Florence. He was a non-drinker in the capital of Chianti, a non-tie-wearer in the land of Armani. He was always smiling. The rest of us gazed with reverence at Italianate grandeur in full figure. Bechir chuckled politely to himself. At the time I thought he was some sort of Muslim Dalai Lama, who had found the secret of inner peace. Upon arrival in Tunisia, I figured it out: he was simply of a people who knew better than to pick olives with their fingers.

Tunisia, along the old Barbary Coast, is a small country of whitewashed splendor, exciting beaches, and mud-walled casbahs. Unlike neighboring Algeria, the French colonized it without a fight and then left peaceably. Tunisia is run by Zine Ben Ali, a hard-minded dictator with a progressive approach to Islam. While Algeria tears itself apart in a civil war pushed by Muslim zealots, German women lounge topless on Tunisia's beaches. Carthage is on a point of land just across from Sicily, not far from the modern capital, Tunis.

The Ben Maads are from Jerba, the island where Odysseus found the Lotus Eaters on his long way home from the Trojan War. By the time Homer's intrepid wanderer landed, the Phoenicians had planted Jerba with olives. Century after century, the trees grew fatter, and their oil went into variously shaped jugs.

Jerba answered to Carthage, to the north, until Hannibal and his elephants dropped in on Rome. The Empire struck back, leveling Carthage in 146 B.C. during the last of the Punic Wars. Jerba began a long history of exporting oil to Italy. Romans covered the island with olive trees. They built a five-mile-long causeway to the mainland, where groves were planted all the way up to the bay of Tunis, a distance of four hundred miles.

Colonial Roman oil merchants earned fortunes and sank them into marbled villas, with baths and gardens. The road from Jerba to Tunis comes suddenly upon the incongruous ruins at El Jem. The ancient city is gone. But a colosseum, not much smaller than Rome's and in better shape, rises above the desert in the middle of a fly-blown little modern town.

In the eighth century, Arab legions moved on to Jerba and beyond, carrying the Prophet Muhammad's holy word. They resettled Tunisia, tending the trees, building casbahs, mosques, and underground oil presses, and they converted everyone but a hearty band of Jerba Jews.

From the 1400s, assorted pirates plundered the coast, followed by Spanish armadas whose storybook forts still dot Jerba's shores. Olives were not much of a priority, just a family business for daily needs. What with periodic scorched-earth invasions from the east and general neglect, the old plantations suffered. Tunisia had about a million trees in 1881, when France came to colonize. Along with their flag, the French planted new groves. Wine, wheat, and olives enriched the territory. By the time Tunisia won its independence in 1956, it had 26 million olive trees. And President Habib Bourguiba had big plans.

Young Tunisia set out to lead the Third World to prosperity: a socialist state of market-minded merchants, a Muslim society which freed women, a one-party democracy with an iron glove on a velvet fist, an Arab-African nation that was neither East nor West, North nor South. Its mineral resources were limited, and wine was not a promising export for Muslims who lived across the water from Italy, France, and Spain. That left olive oil.

Bourguiba planted state groves and exhorted individuals to do the same. Within three decades he more than doubled the number of trees to 55 million. In 1995, nine years after his death, the total surpassed 65 million. Trees cling to bits of clay soil abutting sand dunes on the rim of the Sahara, sharing the horizon with towering palms and the silhouettes of camels headed south. They grow thick in the north, where rains are reliable. They march on for miles within a triangle defined by Sousse and

Sfax on the coast and the ancient holy city of Kairoan in the parched interior.

The Office Nationale de l'Huile, a state-run oil monopoly, bought up each year's crop and exported as much as possible for hard currency. This was a generation ago, before Americans and Europeans laid out large sums of money for artsy little bottles of exotic oils. With earnings from its bulk foreign sales, the ONH imported cheap vegetable oils to sell in local markets at subsidized prices. Olive oil was too valuable for Tunisians to consume and too precious to be left in private hands. Everyone, in theory, shared in the profits. Bourguiba had socialized Tunisia's olives.

Jerba, meanwhile, sat out the oil boom, left alone to languish in splendid abandon. Most new plantations were in the Sahel, the coastal strip from Sousse to Sfax, which included Bourguiba's hometown of Monastir. They replaced Roman groves, which died from neglect or were uprooted by successive waves of marauders. Seedlings were spaced carefully on fields plowed regularly to discourage the slightest blade of grass which might draw away scarce moisture. In Jerba, the old trunks survived on roots dating back to the Phoenicians. Every family owned a few; some had hundreds. In November, in the spring, and over the hot summer, they pruned and pampered as their ancestors had done for millennia.

In the 1980s, however, Jerba was discovered by German tour packagers, Nordic sun seekers, and Club Med. A fringe of hotels sprouted along its beaches, probably the best in the Mediterranean. The word went around fast, and people flocked in by the planeload. It was yet another invasion on an island long inured to outsiders, and its environmental impact was mainly on a strip of coastline known as *"la zone touristique."* But this one threatens to be the death knell for Jerba's magnificent olives.

Bechir's brother Mhenni was waiting for me in his cinder-block showroom stuffed with toilets, bidets, and marble sinks. In his thirties, he owned a thriving construction materials business. Mhenni was smiling; his two other brothers also smiled, as did their father, Ali, who had sold his grocery store in France and moved home to retire. Their good humor, I presumed, had to do with a peaceable life, a healthy diet, and the fact that it was late spring: their cranky, big-time oil mill was shut down for another six months until the next harvest.

The Ben Maads had acted quickly when the ONH gave up its monopoly on exporting olive oil in 1992. They were among a handful of families who obtained permission to find their own buyers at the best price they could negotiate. For the first time since independence, Tunisians had a motivation to produce something better than the minimum standards set by the ONH.

Every winter Mhenni runs the olive press, a quarter-million-dollar continuous chain of grinders and centrifuges built in Italy. The romance aside, it is killer work. When the machinery breaks down, his omnipresent smile is sorely tested by sixty hours without sleep, and he is ankle deep in wrenches. When everything works, the mill pours out forty tons every twenty-four hours, the biggest oilworks in Jerba. By the end of February, Mhenni is back selling building supplies.

There was nothing traditional about Mhenni's mill, but it was clean and efficient, and he bought his olives carefully, in small batches at a time, to keep their acidity low. Because low rainfall means less water in ripe fruit, southern Tunisian olives are plump with oil. Often, the mill could squeeze a liter of oil from three kilos of olives, which was nearly double a normal European yield.

Bechir, meantime, knocked on doors in Italy, hoping to show that his oil would do more for the mass bottlers' blends than anything from Spain or Greece or any other country which might have had a bumper crop. He sought new markets in the Middle East, where importers bought in bulk, and he looked ruefully at all those expensive half-liter flasks on display at the trade fairs he seldom missed. In the spring, when all his oil was sold, he returned to his fruit and vegetable shop outside Paris.

Mhenni was delighted that I'd come to Jerba to see trees. Most foreigners hurry from the airport to the beach, seldom glancing out the bus window at the massive living sculptures which line the narrow roads. They rent taxis to shop for clay pottery at village markets. They crowd into minivans to see a synagogue dating back a thousand years. Only a few notice the olives, and largely because of the tourists, many of the ancient trees are approaching their final years.

"Too many people can make more money doing other things, working in hotels or running a business," Mhenni explained. "Especially if someone has only a few trees, it doesn't pay to do all that hard work for a little bit of oil. You have to love your trees and take care of them for the sheer joy of it. Also, Tunisians travel these days, to the capital, to Italy, to France. Their trees remain at home on the family land, abandoned."

As in Palestine, each time a father dies, his trees are divided among his children. A century ago, a typical farmer might have had a thousand trees, enough to sustain an extended family. Today, the sons of his great-grandchildren might each own a dozen. The Ben Maads own about six hundred, mostly chemlali, a mix of centuries-old beauties and spindly up-starts. But Mhenni buys most of his olives near Zarzis, across the causeway on the mainland, where the French started planting large groves before and after World War II.

Something similar is happening with the date palms, which once soared in splendid stands among the groves of olives. Each winter farmers sawed away the lower branches, leaving a crown of fans atop a tall, slender trunk. Now many are bushy clumps of stunted foliage; dead brown fronds choke fresh leaves that sprout from the ground. The trees grow, but date harvests are poor.

Along almost any road in Jerba, the contrasts are striking. For the most part, dusty, unpruned olive trees grow every which way. The thinning, dry leaves at the top warn that old trunks are near death, sapped of strength by thick shoots growing from the root. And then at an invisible property line, the aspect changes. Sturdy old trees stand in tidy rows, each shaped with an artist's flair around main boughs that produce fresh wood every spring.

Near villages, among shade trees and flowering plants, great handsome old rogues hold their ground. The olives rise high above weathered stone dwellings that were laid out long ago to leave them growing room. Each of these trees reflects a hundred generations of attention, or lack thereof. Some stand high on a single trunk, too fat for three men to embrace. Others look like banyans, supported by a dozen gnarled poles.

When Mhenni first discussed olives, the theme was business. Oil was a commodity, like the plumbing fixtures and tiles that filled his cluttered warehouse. Then we repaired to an outdoor café in Houmt-Souk, Jerba's capital, a warren of stone alleys and arcades that has kept its ancient flavor despite the hordes of pink-cheeked aliens in pedal pushers. The breeze was redolent of jasmine. We drew lazily on bubbling water pipes. All around, stubble-chinned men in grimy wool coats chattered loudly about everything but politics. When Tunisians touch on this last subject, they glance over their shoulders and then whisper.

First we talked about Islam. Mhenni wanted to hear my views. The fundamentalists were a small minority, I said, but they had terrified Americans and Europeans, who seldom spent time looking past headlines. In fact, I had found most Muslims to be generous hosts, respectful of others, appalled by gratuitous crime, and tightly knit by a value system that made sense to their ways of life. Family and tradition seemed to be at the root of their philosophy. I could see, I added, why Muhammad had picked the olive tree as a symbol for his teachings. Did Mhenni know the Koranic verse about the light of Allah?

Mhenni's smile was beatific, and he recited the poetic Arabic in gentle, loving couplets:

"Allah is the light of the heavens and earth, and his light is as a lamp in a niche, shielded by a glass that shines like a star. The lamp is kindled from a blessed tree, an olive of neither the East nor the West, whose luminous oil glows by itself, though no flame is touched to it. It is a light upon a light."

I could not understand the throaty, rolling phrases, but I knew the surah well, from Abdullah Yusuf Ali's annotated edition of the Koran. The lamp, the niche, and the glass make up a triple parable for revelation, spiritual truth, and the transmission of perfect light. Ali explained: "Olive oil's purity is almost like light itself; you may suppose it to be almost light before it is lit. So with spiritual Truth; it illuminates the mind and understanding imperceptibly, almost before the human mind and heart have been consciously touched by it."

Night had fallen by then, and dim bulbs glowed in the trees around us. A scent of grilled lamb wafted across the table, followed by the sharper aroma of fresh fish sizzling in oil. The last tourist buses had gone, and souvenir sellers were bundling up their colorful plates. In the near-dark, I could barely make out the eternal shapes of Houmt-Souk. There was a stone minaret over the mosque, where a muezzin would wail out the next call to prayer an hour before dawn. And there was a giant, twisted tree, whorled and huge at the base, rising to three bent boughs and capped by a spread of foliage that reflected bits of light even after the sun had long gone down.

The next day Mhenni and I crisscrossed the tiny island and explored the adjoining mainland. I admired a seemingly endless grove of well-tended chemlalis. The wide lanes between the rows had been freshly plowed. Mhenni knew only that the trees belonged to the Tunisian government. I later found out who had planted them.

A French friend introduced me to Henri Lamy, in his mid-eighties, who had left Tunisia in 1963. His grandfather had put in the seedlings—sixty thousand of them—in the 1920s. France governed Tunisia as a protectorate, and Lamy's family sunk roots. The grandfather knew nothing about olives back then, but he learned the hard way.

"It was a gamble," Lamy said. "My grandfather was that sort of person, full of initiative. He knew it would be hard growing olives in the desert, but a lot of land was available, and he wanted to try something new."

French agronomists tried to plant intensively, as they did in Europe, but the olive trees resisted. Because of the sandy soil and low rainfall, no more than six or seven trees could be planted per acre. Each tree probed deeply for its underground water. It spread leaves high and wide to catch morning moisture. Over time, the European varieties learned to thrive on less, and then on much less.

After Lamy and other French planters went into full production, Zarzis produced so much oil that concrete gutters carried it through the streets from the mill to the tanker port. When Henri took over with a partner, their extra-virgin oil was sold all over Europe. "One day the Tunisian government took over the groves, and we packed up and went to France," Lamy said. "They gave us nothing. We got an indemnity from the French government, but it did not amount to much. Certainly not what my family had put into those olives over the generations."

The plantations were broken up into smaller holdings, some state-owned and some private. A few are producing better than ever. In Lamy's time, the trees grew a good crop only once in three years; careful pruning has improved that. Others have fallen into ruin.

As we bumped up the back roads, past trees Lamy had planted long ago, Mhenni delivered a nonstop patter on olive lore. "At the start of every harvest near Zarzis," he said, "the women put on their fanciest clothes and all their jewelry and go out to pick olives with the men. They sing songs, make special food. For two or three days, maybe, they dress up, and it is a celebration."

The fun wears off quickly. The trees are massive, and the tops are reached only by tall ladders. Teams of five people rake the branches with goat-horned fingers or, if modern-minded, with specially made plastic claws. Work starts early and ends late, lasting normally from November to February. Pickers are paid by weight. A Tunisian *wiba* equals thirty kilos. And sixteen of those make a *gfiz*. Each one of those earns twenty dinars, a little over twenty dollars. A good crew might pick three to four *gfiz* a day. That works out to about a dollar for fifty pounds of olives, and fifteen dollars daily per worker.

Oil prices are still fixed by the ONH, depending on supply. The 1994–95 crop was crippled by drought, in its second year, and prices rose by 40 percent. Millers earned nearly four dollars a liter, but they spent heavily on olives, electricity, and water. I asked Mhenni if it was a good business, nonetheless. He grinned.

At the time Mhenni had just earned a substantial settlement after a lingering lawsuit against the man who made his press. The catalogue promised forty tons a day and he was getting only thirty-five. During repeated visits, the Italian blamed Tunisian olives for the difference. This, Mhenni countered, was nonsense. The conflict got to be a point of principle. In the end, by tinkering with the process, Mhenni turbocharged the operation beyond all expectations. He considered adding a second unit, but decided that quality would suffer if they turned out too much.

"Producing oil is not a profession with me but an art," he told me. "It is like making love to a woman." Unlike other millers who had expressed this sentiment across the Mediterranean, Mhenni had the delicacy to blush when going into detail. "You have to feel the olive, how it responds, what it needs, what it does not like. If you don't put yourself into the process, your oil cannot be good."

Before I left, Mhenni delivered his *pièce de résistance*. We fishtailed along a faint sand track, barreling through underbrush that closed in from both sides, and took a dozen bewildering turns. Once parked, we walked up a low rocky mound to a narrow opening under a weathered stone lintel. With only cigarette lighters against the darkness, Mhenni led me straight down into a forgotten century.

The underground oil press, dug into the natural rock centuries before was still in use when Mhenni's father was young. A vaulted main chamber was carved around a flat, raised, circular surface in the center. Olives were heaped on the surface and crushed under two granite wheels on an axis. These were turned from dusk to dawn by camels that were blindfolded so they did not get dizzy. The mash was then loaded onto woven mats, those familiar *scourtins*, which were stacked on the press in a cramped second room. Oil was squeezed by a massive beam brought from Alexandria, driven by a wooden screw, and separated from vegetable water by simple gravity. In both rooms, niches in the rock held the original lighting: olive-oil lamps.

Earlier that morning, on my own, I had explored Ali Berbere's cave. It was much larger and older. Instead of a screw press, there was a giant palm trunk on a crude fulcrum; its weight alone squeezed the mats. The mill's subterranean arches were held in place by slabs of Roman marble. A ring of holes allowed families to dump in olives, waiting their turn for camel crushing. Decantation vats were gouged in the stone. It was part of an industrial complex that also turned out pots for storing oil.

But there was something more mystical about Mhenni's cave. It was a perfect example of a workaday part of old Jerba, abandoned like the forlorn trees in a neighboring grove. If you didn't know better, the mound above looked like the sort of vacant lot where you'd dump your garbage. When Mhenni said he wanted to turn it into a small museum, my imagination soared.

As a journalist, I was supposed to shut up and take notes. Getting involved in the process would be like being a scientist who nudges along his experiments. But we were brothers of the olive and I couldn't help myself. I told Mhenni about the Carli museum at Oneglia, how a lot of work and an artful design had brought huge returns. People who did not know they were interested suddenly found themselves passionate about olives.

On Jerba, his idea was full of promise. By adding a brief stop, the hotel tour buses could bring in thousands of summer people, infusing them with the island's heritage. Once people started to look at the old trees, Jerbans would take better care of them. The tourism which threatened to overwhelm the old groves might save them instead.

The side benefit, of course, was all those potential customers for an elegant bottle of Huile de Djerba, should Mhenni bring that into being. Tunisian oil would find its way into kitchens from Stuttgart to Stockholm, and people would want more.

Mhenni listened to me spout out long-range ideas, but I'm not sure what he thought. His response was the Ben Maad version of a poker face: a smile.

Jalloul Jabou, marketing director of the Office Nationale de l'Huile in Tunis, swallowed painfully when I asked about bottled-in-Tunisia oil. A young fireball, with alert eyes and a battered briefcase, he knows more than he wants to about the subject. If Spanish salesmen could barely penetrate the rich American market, what were his chances? Locked out of a European Union that linked Italy, Spain, Greece, and France, where else would he go?

"To sell Tunisian oil, first we have to sell Tunisia," he said, "and our country does not have enough of an international image. Not enough consumers know who we are. We just have to work harder at it."

A shelf behind him displayed what he had to work with. The ONH's brand, Carthage, was presented in a nondescript bottle. It did not matter that its label seemed unlikely to excite impulse buyers; because of small stocks, Carthage was not available for export that year.

Sultan, packaged in Sfax, came in a waxed cardboard brick, like juice or processed milk. It was easy to carry and attractive: an olive-laden sprig against a bright yellow background, with exotic Arabic lettering on the side. The box would keep its contents safe from light. But Sultan's producers were still feeling their way in the domestic market. Customers were not thrilled that some cartons leaked, covering the kitchen shelf in costly oil.

There was not much else. After a few shameless hints that I had yet to taste a typical Tunisian oil, Jabou produced a sample. It was a curious little cut-glass bottle, the top half rounded like the dome of a Turkish mosque, the bottom as squared and thick as a fancy perfume flask. A tiny spout atop was covered with what looked like a plastic green olive. When I poured, oil dribbled down the side. But what eventually reached my mouth was a minor epiphany. This was some of the best oil I had ever tasted. At the prices Tunisians can afford to charge and still make money, it could shake up the olive world. But it probably won't.

Tunisia's oil is complicated business. The predominant tree, the chemlali, is essentially the same variety as the cailletier that produces such good oil on the French Riviera. It suffers no shortage of sun and light, and few pests bother it. Light rainfall on soil rich in nitrogen produces thick bunches of small olives, perfect for pressing. Except during prolonged droughts, the potential could hardly be better.

Tunisians, however, are not great fans of overly fine, low-acidity oil. Partly, this is tradition. If even smoke from a cigarette can flavor oil fresh off the presses, one shudders to think of what happens when it spends days underground in the close proximity of camel droppings. In the old days, farmers stored their olives in sacks for weeks before a donkey lugged them, warm from fermentation, to the nearest mill. Acidity climbed up toward double digits.

Today, in the oil souk at Sfax, people buy a year's supply at a time. It is pumped from clusters of fifty-gallon drums standing in greasy black puddles. Reused containers for sale bear their original familiar markings: BP, Esso, Shell. The oil is dark and pungent, thick enough to lubricate a piston, and most buyers would not have it any other way.

Recent history has not helped. Bourguiba's grand scheme worked no more effectively than most experiments at state socialism. Farmers earned so much per *wiba* of olives, and quality was not broken down into nuances. Oil was bought by weight, only by the ONH, at prices fixed in Tunis. Batches were mixed together in containers to be sold by the ton. In a business driven by personal commitment, incentive played little part. Vagaries of the olive trade produced occasional surpluses which ended up aging in storage tanks.

Under a new law, producers like the Ben Maads are doing things differently. The best Tunisian oil is fresh and clean, as extra virgin as anyone else's. Picked in November, it has a slight nip, without the afterbite of greener European oils. But there is that image problem, which makes Jabou so miserable.

"He is a liar," Jabou said, when I mentioned that a major Italian blender had told me he routinely rejected Tunisian oil as too coarse. Sputtered would be closer to it. "I did the contracts myself. He bad-mouths our product and yet 30 percent of the oil he sells is from Tunisia. A liar."

Accords with the European Union allow Tunisia to sell 46,000 tons a year at lower duties. After that, it is war, and in good years the Tunisians have another 100,000 tons to sell.

"Italy, especially, puts up artificial barriers which freeze us out," Jabou said. "We complain and nothing happens." By International Olive Oil Council guidelines, oil is graded by acidity levels and its properties of taste and aroma. But any country can add its own standards. "The Italians study chemical analyses of their oils and ours, and then they fix levels for certain components which disqualify us."

There was, for instance, the K232 factor, a laboratory measure of how quickly oil mash oxidizes. European Union guidelines put the limit at 2.5, just above the normal level for European oils. Tunisian oil is usually a bit higher. "This has nothing to do with taste or quality or nutrition or value," Jabou concluded. "It is only a technical trick to eliminate the competition."

I checked this with Alissa Mattei at Carapelli in Florence. A high K232 rating suggests instability, she said, but slight differences below and above 2.5 mean very little. Hans-Jochen Fiebig, a German chemist who tests oil for the European Union, agreed. "They pay so much for subsidies and price support that I suppose they want to have a lot of regulation," he said of the Brussels bureaucracy. "Every year there is more, and much of it is not really necessary."

Beyond Europe, Tunisian exporters face the same obstacles as the Spanish, at a much greater disadvantage. Although Mafia influence on

American imports has dwindled over recent years, human nature and the old-boy network have not. If the central buyer for a supermarket or a restaurant chain is Italian, he is likely to lean toward what—or who—is familiar. And for consumers, Italy represents a culture of artful eating. Buyers might be comfortable with Spanish or Greek oil if the price and the presentation are right. But Tunisian?

Under normal circumstances, Tunisians would act like the Greeks and eat what they don't sell. But again, traces of Bourguiba have altered habits. Because of heavy subsidies, packaged olive oil, like Sultan, costs six times as much as generic vegetable oil on supermarket shelves.

What started as a scheme to free olive oil for export has grown to be an institution. Tunisians have gotten used to frying food in cheap soya and seed oil. The ONH would love to eliminate the costly subsidies, for which no economic reason remains. But there is a simple enough political reason. A sharp price rise might send people howling into the streets. For the tough but pragmatic leaders of the country's only real party, nothing is more important than staying in power. Not even olive oil.

Leaving the Office Nationale de l'Huile can be a lengthy process. The patch of garden outside commands an appreciative pause. Among the olive's most attractive qualities is the company it keeps. This can range from high pines to scraggly cactus. And in northern Tunisia the spread is about as wide as it gets. Outside the OHN rich red and pink hibiscus nestle among palm trees, yuccas, and junipers. Lush oleander blooms are surrounded by delicate flowers. The centerpiece, of course, is a perfectly pruned, magnificent olive.

As it happened, I was in a hurry. The market would be closing soon, and I had olives to eat. However North Africans make oil, their cured olives are seldom surpassed. After years of bragging about the local wares at the Draguignan market, I learned the hard truth: some of my favorites were imported from Morocco and Tunisia.

A brisk walk up past the flower sellers and newsstands along the broad promenade of the avenue Habib Bourguiba takes you past the iron gates of the French Embassy and on to rue Charles de Gaulle. Below is the Central Market. I skipped the fresh fish, the mutton carcasses on hooks, and the mounds of exotic fruit, pushing on to find Youssef Bribani standing among his barrels of aromatic olives. Slight and balding at thirty-five, he was third in a line of master curers.

I asked him if his were the best in the market.

"It is very difficult to use the word 'best,'" he replied—which is Tunisian for yes.

Bribani did a brisk business in saheli, a general name for small spicy black chemlali and meski olives. But his specialty was the violette, a fat purple chaal that grew in the north. The olives are placed in a barrel with every layer covered in salt. After a month or so, their natural water absorbs the salt, but the flesh is almost sweet.

"Normally these olives are very bitter, and we must treat them," Bribani said.

"How?" I asked.

"With a wild grain we find in the mountains."

I pressed for details. With an evasive smile, he gently changed the subject.

"Olives mean everything to the Tunisian," he said. "If you have trees, your children are provided for. There might be famine, war. As long as you have olives, you will always eat."

Driving around the country, you realize that the olive has been man's best friend in most of Tunisia for at least three thousand years. And in the ancient ruins of Dougga, an hour's drive west of Tunis, the sense of history is overpowering.

What remains of Dougga is mostly Roman, a grand complex of amphitheaters, temples, and baths, surprisingly well preserved. For lovers of archaeology, the colonnaded *capitolio* is probably the most dramatic trace of Rome left in North Africa. I marveled at the old rocks but could not help focusing on what continued to grow between them.

Seedlings that had bordered the buildings in Roman times were now great monster trees, and their roots reached deep into stone foundations. No one, apparently, had ever thought to stop the encroachment on them. These days, the overall effect is thrilling. Huge hedges of prickly-pear cactus—*les figues de Barbarie*—spray yellow blossoms as high as the olive branches. A sort of mutant hollyhock rises like Jack's beanstalk, adding bright pinks and purples. Oleander clumps blaze in deep red.

Many of the olives are in shabby shape, neglected by curators with other priorities in mind. I mentioned this to Mohammed Salah, a graying old guard in a frayed sweater, and his brow furrowed. He had had the same thought: his own twenty-eight trees were in far better condition. Salah took my arm and directed me through a labyrinth toward a path that led to the plain below. A stone Punic mausoleum stood alone in a grove of ageless trees.

Down below, I flopped onto the grass beneath a waist-thick branch and thought about the guide at the Garden of Gethsemane who had told me the trees there were five thousand years old.

In the peaceful air, I studied the Punic tower and a giant tree next to it. It was hard to work out which was older. But each year the wind and sun wore away more of the stone mausoleum, which had lost its purpose two thousand years ago. The tree, putting out new growth with every spring, just got bigger and bigger.

Dar El Jeld restaurant, just past the prime minister's office at the fancy end of the casbah, is an earthly version of an Arab-style paradise. From the outside, it is a whitewashed urban fortress, undistinguished except for its massive, bronze-studded, carved wooden doors. Initiated patrons slam the heavy metal of the ornate knocker and a jinni appears in a black tuxedo.

The soft light inside glints off brass fixtures and illuminates walls tiled in richly colored patterns. The beams and carved doorways are in handsome old wood. The floors are an elaborate mosaic, worn over generations to an exotic patina. Lush tapestries and plush cushions suggest serious postprandial lounging. Servants with tongs walk around dropping fresh coals into braziers to warm the room. But at Dar El Jeld one's attention fixes on the menu. Tunisian cooking is not all couscous.

"My brother and I loved the cuisine we grew up with, and we thought it would be fun to try a restaurant," Habiba Abdelkefi told me in a comfortable upstairs office where she felt right at home. The eighteenth-century mansion was, in fact, her family's house. Habiba is Parisian chic, with gracious Arab manners. Her brother, Ali, is impeccably tailored, with the smooth charm of the well-born. They are natural restaurateurs.

"Neither of us had any experience," she said, "but my friend likes to entertain. Now she is the head chef." That was in 1989. Now all hundred places fill every night for meals at substantial prices. "Tunisians like to bring their foreign guests here. We served the King of Spain. His ambassador preferred to receive him here rather than at the embassy."

Dar El Jeld's specialties are lamb and fish cooked in elaborate but delicate sauces, variations on old Tunisian themes. Raisins, lemons, and piñon nuts blend with sharp spices. Fresh sea bass is poached in tomatoes, onions, and lemon *confit*. Tender meat, roasted and lightly stewed, falls apart at a fork's touch.

And the common ingredient is thick, yellow Tunisian oil. "All our dishes are based on olive oil, starting with couscous," Habiba said. "It is the heart of our cuisine."

Not for everyone. With economic hard times, pressure increases for Tunisia to sell more olive oil abroad and push seed oil at home. Run-of-the-mill restaurants cut corners. Workers wolf down plates of soggy rice and meat, or sandwiches of canned tuna and green peppers in vinegar. Housewives scrimp during the year to splurge on holidays. At best, most buy an ONH blend of olive and soya oil. But it has been that way, more or less, since ancient Carthage and Rome. Olive oil is for the peasants who make it and the lucky ones who can afford to buy it.

After a lunch to stuff a pasha, I asked to speak with the chef. She was out, and I settled for a mint tea and honeyed pastries baked with olive oil. It was just as well. I was not sure how to handle a recipe that begins: Install a heavy brass knocker on your front door.

In Sfax, I found some scientific answers to questions that I had almost given up asking. Ahmed Trigui at the Institut de l'Olivier spent half a day educating me. He was no amateur. His most recent paper explored how olive leaves shaped themselves to accommodate whatever the seasons dealt them: tiny hairs around the pores on their undersides monitor the weather; the leaves curl inward to slow transpiration in dry spells and open flat when soil is moist. That is why olive trees are silver—leaf undersides showing—in pleasant weather and leathery green when the sun beats down.

Ahmed had studied every aspect of the plant's biology and its place in society, and he was finishing a history of olives in Tunisia. What he did not know was somewhere in the papers towering behind him. And each day more data accumulated on some olive node of the Internet.

When Ahmed paused for breath, his colleague Taieb Jardak offered elaborate CVs on every pest that had ever penetrated an olive grove. Their scientific insight into the plant's ecology explained why Columella's "queen of trees" so easily turned perfectly normal city folk into hoe-wielding olive zealots. To raise healthy fruit, you must talk to your trees in a way that goes far beyond a few cheery words to the ficus on the terrace.

Back at Wild Olives, I had fought dubious battle against the trees' enemies, one by one. Each insect was a familiar foe, with markings as distinct as gang colors. Dacus, the olive fly, looks like a bug-eyed housefly with three stripes on its chest and lace-patterned wings. Coccus, the black scale, has an *H* on her back; she looks like a peppercorn affixed to a twig. Our *teigne d'oliver* was, I knew, a *Prays oleae*, a moth that eats the buds and flowers. The thrips, *Liothrips oleae*, were flying insects too small to see well, but they fouled olives and curled leaves. The neiroun, *Phloetribus scarabaeoides*, was a little black beetle that bored into wood.

Past research had suggested tactics ranging from helpful to hysterical. L. Degrully's *L'Olivier*, published in 1907, told how to combat the *Otiorhynclus meridionalis*, which hid underground by day and ate leaves after dark: Sneak up in the dead of night—no noise, no light—and spread a white cloth under the tree; they'll drop onto it. Eight decades later, the International Olive Oil Council's 350-page manual on olive pests (only insects; diseases took another volume) advised not even bothering to fight.

Poisoned traps and judicious spraying could control the fly. Copper spray in spring and fall held off black scale, as well as a leaf disease known as peacock spot. Bacterial infections, such as one called olive tuberculosis, which caused ugly tumors, and another known as olive leprosy, were controlled mainly by good fortune. The only practical counsel was not to use infected shears on healthy trees.

It turns out, Ahmed and Taieb Jardak explained, that nature has answers for these plagues. Olives need room to grow, air and warm sunlight in their branches, and aerated soil around their roots. Once a tree is healthy and pests cannot lurk in nearby vegetation, they resist almost everything. Tunisia is a laboratory case. Where rain is scarce, farmers plant as few as eight trees to an acre and leave only bare plowed earth between them. Thus separated in the hot sun, Taieb's cast of characters seldom bothers the trees. Only 3 million of the country's 65 million trees are sprayed for pests. In Europe, underbrush grows fast and neighbors are often lackadaisical about pest control. But where care is practiced, olives flourish.

In each region of every olive-growing country, nuances in soil conditions and weather patterns affect the fruit, Ahmed said. And olivemen's customs vary accordingly from place to place.

Tunisians seldom bother to graft their trees, letting nature work its will. They have learned how quickly olives adapt to new circumstances. Normally, olive roots grow sideways; south of Sfax, taproots can reach down six meters. Varieties evolve over time. The picholines were transplanted originally from France, where they need more than 300 millimeters of rainfall a year. In Tunisia they now settle for 200 millimeters, and they grow faster than French picholines. A giant chemlali might produce 800 kilos of olives. Its close relative, the French cailletier, seldom approaches 50 kilos.

Most Tunisian trees are chemlali, zalmati, chaal, chetoui, and meski. The barouni, found widely in California, came from Tunisia. Altogether, Ahmed believes, there are fifty or even a hundred other varieties around the country. Most are self-fertilizing, but some require a different variety nearby to produce the pollen they need to flower. And scientists are seldom sure how the trees cross-pollinate. After eight thousand years of olive growing, much remains a mystery.

"Olive research is a very slow process," Ahmed said. "If you want to be famous or wealthy, you do something else. A great scientist in Montpellier, *le Père* (M. A.) Bouat, worked all his life on the fertilization of olives. He was never able to determine an effective dose for any variety. There is not one olive that resembles another."

I asked Ahmed about the lifespan of olives. Andrés Núñez de Prado cut down his trees at eighty years. Those old giants at Dougga, probably headed into their third millennium, were still growing olives. The answer had to lie somewhere in between.

"These trees don't come with birth certificates, and we can only guess," he said. A healthy olive trunk throws out fresh buds for centuries. Because roots expanded and grew new trees, dates were impossible to pin down. In the end, Ahmed the scientist supported others' common wisdom: "You can say that olive trees are immortal."

Trees produce far longer if the old wood is cut away, he said. But except for a few basic principles, there is no right or wrong way to prune; it is an undefinable art. It is true, he said, that whacking the tree damages young wood. Still, it causes no permanent harm. Like the continuous system mill, it has its economic advantages over the old way. Ahmed did not sneer at mechanical pickers, as I had expected. "The roots underground are just like the limbs above," he said. "If you shake them a little once in a while, it does them good."

Beyond his knowledge of the science of the olive, Ahmed loved the taste and the lore of the noble little drupe which had helped civilize the world. Here, I realized, was the perfect arbiter for the great debate on the best way to make oil.

Ahmed laughed when I told him of the fight-to-the-death convictions I had heard, in every variation. His short answer was that the old way was best, pressing slowly in mats and then separating the oil from its water by lifting it off with a scoop or letting it decant in linked chambers. Centrifuges could shock the oil, and extra water flushed out its flavor. But then he added provisos.

If an inexpert miller dipped too deeply, vegetable water remained in the oil. If he did not carefully leave behind all the sediment, impurities could turn rancid and infuse the batch. If the mats on the press were not perfectly clean, the oil would pick up an odor. If any strong smells wafted through the mill—chemicals or tobacco, for instance—the oil would pick them up. "Olive oil is so sensitive," he said, "that if it is shipped in containers aboard a vessel that has carried petroleum or phosphorus, you'll taste it."

In practical terms, he continued, old-style pressing was for small producers and purists. Centrifuges were hard to avoid. The best thing, as Armando had said in Lucca, was to use the new machines that spun at slow speeds and recycled the olives' vegetable water for the separation process. Adding heated water was the scourge of oil making.

"Water washes the oil, carrying away elements of flavor, and water over 25 degrees destabilizes it," he said. "They talk about 'cold-pressed' oil, but the water is almost always too hot. I go to mills, and they tell me the temperature is no more than 35 degrees [95 degrees F]. But I stick my finger in the oil, and it is hot."

To make this point, he held a clear flask up to the light. The oil inside was a rich dark yellow. "This is two-year-old chemlali," Ahmed said. "If this had been washed with warm water, you would not see anything close to this color."

What tastes best, in the end, has less to do with science than with cultural preferences. In some remote Tunisian villages, farmers dry out their olives and store them. When the oil runs low, they take a sackful, and their camel, to a communal mill not much different from the one Mhenni showed me in Jerba. Someone finds the key to open the mill, and they make a new batch. Within a few months after harvest, the oil's acidity might be 16 percent and, later, as much as 30 percent. In other words, nearly ten times higher than the international norm for sending olive oil to the soapmakers.

For good measure, Ahmed took me to a soap manufacturer. Mohammed Chaari ran a bustling factory outside Sfax, which treated pomace, the dry leavings of first-press oil. The pomace was trucked in and piled up in mountains. If it is stored long enough, the acidity will approach 40 percent, and oil from it will go directly into soap.

If it is fresher, however, the acidity might be in the range of 10 percent. With the high heat from steam, Chaari could coax six and a half liters of pomace oil from a hundred kilos. The process is complicated and smelly. One chemical bath adds a lye compound to neutralize the oil and bring down the acidity. Another filters it through activated earth to remove its dark color. Then it is deodorized with a different dose of chemicals.

The procedure for making "pure" olive oil is not much different, except that the original ingredient is inferior oil milled from olives rather than coaxed from olive waste. It still must be refined with heat and chemicals. Before packaging, some extra-virgin oil may be mixed in. Its final acidity may be under 1 percent, but it is a washout on organoleptic ratings. Pure olive oil and pomace oil are perfectly good for frying or foot massages. It is best, however, not to watch them being made.

———

After tea at an ancient café atop the city walls, we slipped and slid on the greasy black cobblestones of the oil souk. It looked less like a market than a busy truck-stop service station. By mid-afternoon, Ahmed had immersed me in every aspect of olivery, but he had one more item on the agenda.

I plowed after my generous mentor through the crowded, narrow lanes of the Sfax medina, turning a hard left at the vegetable stands and pushing on into the deepest part of the old stone casbah. He called an abrupt halt at a tiny shop where two aging men poked and scraped with crude tools. Their handiwork was hung up on strings, brownish curved claws like rudimentary castanets. Ahmed did not want me to leave Tunisia without my own set of olive-picking goat horns.

Moroccan Olives

Patala M'Qualli

This lamb-and-potato dish features the yellow sauce of oil, ginger, and saffron that Moroccans call *m'qualli* and juicy purple olives. *Cooking With Olives*, from the International Olive Oil Council, offers this recipe.

2 pounds lamb cut in pieces
7 tablespoons olive oil
2 onions, finely chopped
3 garlic cloves, chopped
1 teaspoon ginger
1 tablespoon fresh coriander
  Salt
1 pinch saffron

2 pounds small [yellow] potatoes, peeled and cut in pieces
3 tomatoes, diced
7 ounces pitted purple olives, Moroccan or Tunisian
1 tablespoon parsley, chopped

Place the lamb in a large pan with the oil, onions, garlic, ginger, and coriander. Cover with water, add a little salt and the saffron, cover, and cook about 1 hour. Add the potatoes, tomatoes, olives, and parsley. Adjust for salt and cook another 15 to 20 minutes, stirring gently while trying to keep the potatoes from breaking up. Transfer the meat to the center of a serving dish and arrange the potatoes around it in the shape of a star. Pour the sauce over the center.

Serves 4.

"Make me poor in wood,
I'll make you rich in oil."

"Caress me, don't beat me,
If you want my fruits another time."

"Prune me hard, fertilize me well,
If not, leave me for another."
—Moroccan proverbs; the
olive tree is speaking

10

Marrakesh

Centuries before Christ, rich trees grew near Volubilis, at the northeastern edge of the Sahara, a far-flung outpost of Hellenistic culture. Roman conquerors built a greater city, near where Meknes now stands. As in Tunisia, the Romans planted olives all the way from the Mediterranean coast. These days Moroccans grow fewer trees than Tunisians, and they export little oil. But when it comes to olives for eating, they are the world champions.

At any market in Morocco, tubs are heaped with olives: black ones cured dry in salt or soaked in brine; green ones, fiery with red chilies or sweet with lemon *confit*; violet ones, plump and firm in garlic, parsley, turmeric, oregano, and rosemary or thyme. Every village has its secret preparation, and each city has its specialty. In Fez, they add coriander. In Agadir, on the Atlantic coast, they sprinkle in secret herbs from up in the mountains. Olives sneak into every sort of dish, from chicken *tajines* to refreshing salads.

For olive lovers, there is no place on earth like the little corner of the Marrakesh souks, just past the cobra charmers and pickpockets of Jemaa el Fna Square. Much of Morocco is olive country, from the coastal fringe, up the slopes of the Atlas and Rif ranges, and on toward the Sahara. Trees grow most densely near Fez, a thrilling thousand-year-old imperial city of high walls and winding lanes. But Marrakesh is a holy site in olivedom.

La Mamounia, an Art Deco palace which is many travelers' favorite hotel anywhere, was built in the 1920s in an olive grove within the ancient imperial capital. Huge trees stand in a lush garden among oranges, succulents, and delicate blossoms. Flowering vines climb discreetly up old trunks, pruned so that the silver-green crowns seem to sprout crimson blooms. Marrakeshi olives are prepared in cellars along the back alleys, and Berbers come down from the Atlas Mountains with their own specialties.

I found the olive souk, wending among tourists, beggars, and assorted guides who claimed direct descent from the Prophet. It was not far from where auctioneers sold slaves until 1912, when the French took control of Morocco. A half dozen stalls, similar except for the graceful Arabic script announcing their ownership, displayed overflowing barrels and bins.

Each stall was framed on three sides by high walls of glass jars, all arranged with the elaborate care of a tile mosaic. Brilliant yellow marinated lemons alternated with scarlet peppers against a dozen shades of green. In some jars, olives were set geometrically among small pointed chilies. Others contained olives with white almond chips. Mixed olives nestled in pickled carrots and celery. Sliced green olives were artfully packed in patterns of little wheels.

On broad counters, in side-by-side enameled pans, a half dozen perfectly shaped cones of olives rose two feet high. In each stall a Central Casting olive seller clutched a large scoop, ready to fill his waiting plastic sacks.

Under a sign reading ELBASSIRI ABDESSAMAD, COMMERÇANT D'OL-IVES EN GROS ET DÉTAIL, I found a friendly oliveman, mustachioed, in a grimy shift and white skullcap. His name was Ali. Until four years ago Ali worked at the stall across the lane, selling ladies' undergarments. He likes olives better.

With courtly patience, Ali described his staples. Almost every olive in the country is a Moroccan picholine, or beldi, a hardier, wilder variety than its French cousin. His were no exception. The green olives are picked from mid-September through October. They are cracked and soaked in salted water, often with caustic soda; some are pitted. The violets come next. These are sliced once with a razor so brine can penetrate. The blacks stay on the tree until late November and December, ripened enough to be cured without breaking the skin.

He was pushing the citrus-laced olives, fat green ones mixed with bitter orange and lemon peel soaked to sweetness. But I zeroed in on a tub of blacks, greens, and violets that were smothered in lethal-looking peppers and parsley flakes. How long, I asked, had those been curing? Ali looked puzzled. I explained how my neighbors in France did their flavored olives; aromatic herbs, hot peppers, and spices sat in the mix for months so that all the flavors blended together. He laughed. "Watch," he said.

Ali took a large empty pan and scooped in piles of olives *au naturel* from their various bins. He peeled a few large cloves of garlic and sliced them. Next, he sprinkled chopped parsley liberally over the olives. Then he scooped in enough chili paste to paralyze a Mexican bandit. With a final flourish, he poured in a few respectable glugs of oil. Ali seized the pan and shook it vigorously, tossing the olives into the air with a practiced flourish. He had made ten pounds of Marrakesh olives in four minutes flat. They were fabulous.

It is not always so simple. Subtle flavors are better if they have a long time to infuse. Olives in fennel, for instance, take months. If olives are to absorb the taste of whole chilies, they should steep for a week or so. But take it from the Moroccans: one need not be neurotic about flavoring olives. All it takes is naturally cured olives. With a little imagination, personalizing them is as easy as tossing a salad.

The next morning at Elbassiri's factory in a dusty slum at the edge of the Marrakesh medina, I saw the other end of the business. Six women work from 8 a.m. to 7 p.m. for a little over two dollars a day—the price of four pounds of olives in the market—with no paid time off and not even a free aspirin for a health plan. Halima Snani, the woman in charge, seemed cheerful enough about it all. Her teenaged sister also worked the olives, and they made enough to feed her two kids and six others in their tiny house. Moroccan unemployment approaches 40 percent, and olive curers were not in short supply.

The plant is an open courtyard filled with blue plastic barrels. Halima, a birdlike woman with beautiful eyes and horrid teeth, manhandles hundredweights as if they were handbags. In one corner leaking sacks of caustic soda are piled high, each labeled DANGEROUS SUBSTANCE: ATTENTION! CORROSIVE. The olive women use gloves to handle them. In another corner there are sacks of rock salt. Barrels froth with evil-looking foam. Hoses slop water into the leaching tanks.

Halima showed me the primitive hand-lever pitting machine. She demonstrated how the nicked old Exacto knife sliced violet olives but not, Allah willing, the supporting finger beneath. She smacked a green olive with a wooden paddle on the stone ground to show the cracking process. Halima waited for my next question, but I did not hang around to ask it. I enjoyed eating Moroccan-style olives far too much to learn more about how they were made.

On a grander scale, packing olives is a major industry. In a good year Morocco exports 60,000 tons, second only to Spain. Large companies sell California-style blacks, cured in lye, and Spanish-style greens in brine. Moroccans supply three-quarters of France's olives, many of which are then sold abroad as French. The Spanish are beginning to buy in bulk. At a dollar an hour, labor costs thirteen times less than in Spain. It is cheaper for the Spanish to haul olives across the Strait of Gibraltar and re-export them.

The Moroccan picholine is equally good for oil or the table. When the crop is meager, competition is fierce. A handful of big mills bid against the packers. Rural growers, anxious to supply their own needs, are reluctant to sell. Other obstacles plague the olive industry. I stopped in Casablanca to see Don Humpal, a Californian who did Peace Corps service in Senegal, who now runs a U.S. foreign-aid project called AMI—Agribusiness Marketing Investissement, in that French-English *lingua franca* of Franglais. The idea is to help Moroccans pack and sell olives abroad.

Humpal was optimistic, but not wildly so. The biggest potential market is the United States, but Americans like their own canned and pitted bland olives. Tariffs protect California growers. Moroccan hygiene leans toward lax, and marketing and cost-analysis systems are also unsophisticated. An AMI report noted: "Moroccan firms have a reputation for taking orders and not delivering the product ordered or delivering it only with an additional financial assistance from the importer."

But the government is pushing hard. Each year at least a million seedlings are added to Morocco's 50 million trees. Some are for windbreaks and soil conservation, but most are for table olives. The problem, however, goes beyond trees. As far as olives are concerned, Humpal said, rural Morocco has not changed much since the Middle Ages. There are new roads and fancy communications, but people like to do things the old way, and few have much incentive to do otherwise.

A few large mills produce respectable extra virgin, but most oil is made in sixteen thousand *mâasras*, small old-style mat presses, where donkeys or camels power the grinding stones. Oil is dark and strong, the way Moroccans like it. And old ways prevail with table olives. Besides the packing plants, nothing is consistent about artisanal olives. They may taste wonderful, but no two batches are the same. Humpal laughed when I asked how many different preparations I might find. He replied, "How many households are there?"

At the market in Fez, late in September, Ahmed Senhaja sat amid his baskets of green olives fresh off the tree, the raw material for do-it-yourself curers. Most of his customers would treat them the way Palestinians do: crush the olives, soak them for eight or nine days while changing the water regularly, and then add salt. The salt not only removes the last bitterness but also wards off fungus and other unwelcome cultures. Hot green pepper usually ends up in the mix.

Senhaja is a Berber from the Atlas Mountains who has been growing olives for most of his eighty-two years. He did not know what to make of my interest in the subject. I asked if Morocco's savage drought, then in its second year, had affected his trees. He cackled. "If the rains come, or they don't come, the trees don't care," he said. "They just grow."

Like a lot of his neighbors, Senhaja hardly touched the trees his great-grandfather had planted. They grew at nature's whim. Each harvest time, the high branches got farther beyond easy reach. Senhaja's grandchildren just got a longer stick.

"I leave them be," he said, cackling again. "We pick, they grow." He swept his hand skyward to be sure I was following his drift. "Olive trees are wonderful things."

That explained what I had seen on the drive up from the coast: huge, bushy trees that hardly looked like olives. Massive tangles of branches grew helter-skelter from twisted main boughs. Brown twigs fanned from deadwood. To a Tuscan traditionalist, this would have triggered cardiac arrest. But Senhaja's olives were large, smooth, and healthy, free of olive-fly holes. Someone was doing something right.

A few other olive sellers squatted on their haunches near Senhaja. The tubs of cured olives were scattered in shops down narrow alleys. For all its tumult, Marrakesh has a certain order, with oases of breathing space and horse-drawn cabs to whisk away the weary visitor in style. In the market, each craft and product is clustered in its own souk.

Fez is mere bedlam. It may be the world's largest living medieval city. Cobblestone lanes, crammed within the massive walls, veer suddenly into unexpected directions or drop at a sharp angle. A distracted visitor, though safe from cars, is likely to get gored by a passing donkey saddle.

Fez is from some other age, so far out of time that the American Express decals and tourists in tennis shoes offer no reality check. It was Idriss II's capital in 808, the heart of the first great Moroccan kingdom. Fez el Bali, old Fez, was built up on two banks of a dry *wadi*. On one side, Moors evicted from Spain brought back an Andalusian flavor. On the other, settlers from Kairouan added touches of Holy Land Islam. African treasures came north on camelback. Italianate grandeur came by sea. Together, it is a blood-pulsing mosaic of ceramic tiles, vaulted archways, carved precious woods, soaring mud minarets, and polished stone lanes that drop away and disappear in the dark.

But in Fez the impact is not so much from the place as the faces, each half hidden by exotic headgear. A wild black mustache bristles under the pointed hood sewn onto a coat of many colors. A crocheted white skullcap sets off burnished copper skin. A woman's black veil masks all but voluptuous dark eyes. Olive trees often differ drastically from culture to culture. And so do olive people.

Nestled in the valleys of the Rif Mountains, Fez and nearby Meknes are the heart of Morocco's olive country. Small tribes scattered in the hills tend haphazardly placed trees. A few old families keep large groves. In Meknes I asked for a briefing from Moha Marghi and Ahmed Oumekloul. Marghi is chief government agronomist for the Meknes region. Oumekloul is the director. Both wore a suit and tie, but each was a teenager before learning to use a fork. They were worried about Morocco's olives.

"It is the rural exodus," Marghi said. "Too many young people are going to the city. They want the benefits: cafés, electricity, a good time." For him, the irony was bitter. Around Meknes unemployment reaches 50 percent. Young men are so desperate for work that they risk their lives in fragile boats to slip illegally into Spain. But the 40,000 acres of olive trees near Meknes require 800,000 workdays a year, and many go untended.

"Even young people who want to stay find that they can barely survive," Marghi said. "They have no land; plots are getting too small. No one can make it on only olives, and they cannot grow enough of other things."

Here was that problem again. When a man dies, his land is divided among his sons, and this has gone on for too many generations. Farmers cannot make it without at least thirty acres, Marghi reckoned, and now the average inheritance is less than that. Families can make arrangements among themselves, but land is only one of the problems.

Successive years of drought have weakened the trees. In 1994 rainfall in Meknes that should have measured 500 millimeters was only 280, the lowest amount in 140 years. And the government ended subsidies to olive growers in 1987. In those years, the state guaranteed minimum earnings and even helped with the cost of fertilizer.

Now the state gives trees. Each year, Marghi said, the Meknes nurseries turn out hundreds of thousands of olive seedlings to be trucked all around the country. Many farmers had big plans. More production might bring down the price of olive oil, a blessed development for most Moroccan families.

Olive oil is so expensive that many housewives use seed oil as a regrettable alternative. As in Tunisia, people like their olive oil strong, up to 5 percent acidity. This is not only for the throat-burning taste, but also because only a little bit flavors a whole dish. A liter of good olive oil might cost thirty-six dirham, about four dollars. Seed oil sells for ten dirham. "Families eat olive oil on little plates at breakfast," Marghi said. "Many would like to cook with it, but they can't."

Fortunately, we all repaired to the home of Ahmed Nidaoui, a Meknes journalist whose wife, Yassef Fouzia, did not skimp on the oil. She had prepared a traditional Friday couscous. Facts and figures gave way to the reminiscences of men who treasured their olives. As we talked, we dug into the couscous—literally. Morocco-style, diners ram their right hands into the heaping plate, palms upward, to extract a handful of semolina and bits of meat and vegetables steeped in spicy sauce. This is rolled into a moist ball twixt fingers and thumb and popped into the mouth. If this does not sound easy, it is harder than it sounds.

Juice dribbled down my wrist and couscous grains paved my chin. No one cared, if anyone even noticed. My hosts were busy surveying the fast-diminishing pile for likely bits of lamb or chicken and succulent chunks of yellow squash and sweet turnips. A few choice items were shunted my way, for politeness. Others disappeared at the speed of light. The key to it, of course, was the perfect texture of homemade couscous, so light the grains nearly floated above the platter. For this, North African cooks insist, you need olive oil.

Tunisians like their couscous savory but simple, with lamb, chicken, onions, garlic, potatoes, carrots, zucchini, turnips, and hot peppers. Moroccans add touches of the sublime: cinnamon, raisins, pine nuts, ginger, cumin, coriander, saffron, thyme, and tarragon. And that is a mere beginning. At the far western end of the Arab world, Morocco blends a dozen cultures in its cuisine: Arab, Berber, Jewish, Andalusian, French, and different shades of Africa and the Middle East.

In Morocco, cooks use every spice that ever found its way across the desert or the Mediterranean, starting with their own beloved *Ras al Hanout*, a powder of thirteen peppers, berries, herbs, and spices. Their oil is strong enough to stand up to honey and candied citrus. Fragrant waters add subtle scents.

Had it not been Friday, we might have had *bisteeya* (or *bastela*, among a half dozen other transliterations), best described by Nancy Harmon Jenkins: "A rich fantasia of minced pigeon and almonds, butter, sugar, and eggs, all layered between crisp, transparent leaves of *warka* pastry and flavored with a complex of pepper and cinnamon, ginger, cilantro, saffron and orange-flower water." It could been have a lush *tajine* of meat braised with prunes. This being Meknes, it more likely would have been flavored with olives, added toward the end to preserve their crispness. No matter. The couscous was terrific.

At Volubilis, ancient Berbers tamed a few wild oleasters for oil. Phoenicians moved in and improved the production. But it was the Romans who built the vast stone and marble city, a walled and colonnaded imperial outpost capital that was all but devoted to olive oil.

Volubilis was the chief inland city of Maurétania Tingitana, the neighboring Roman province to Maurétania Caesariana, which took in Algeria. It flourished under Hadrian, the olive-loving emperor who planted trees from the coast to the dunes, all across North Africa. In A.D. 285 it was destroyed by the Vandals, who cut down the trees to discourage any phoenix-like revival. The few remaining buildings stayed largely abandoned for a millennium. Moroccan rulers swiped much of the marble to build Meknes, and the 1735 earthquake that leveled Lisbon did the rest.

Archaeologists have uncovered only a fraction of Volubilis, but they have already come across sixty public olive presses, not counting the scores of others in private villas. An imposing restored stone structure in the main square looks as if it could have been the city's treasury; it was actually an oil mill. The Romans used mushroom-shaped granite wheels to crush the olives in the manner of an inverted mortar and pestle. Mash was stacked onto mats. Instead of stone weights, as in Ekron, Volubilis used slaves. Two burly men would crawl through a small opening that was bolted behind them. Then, all day long, they would load the mats onto the press and use their combined weight on a long beam to squeeze out oil.

I was taken in hand by Hamed Charradi, a twenty-five-year-old archaeology student who hung around to offer his services to the occasional passerby. He pointed out what has been lovingly excavated. The triumphal Arch of Caracalla leads to the remains of the main western gate. An elaborate brothel is easy enough to identify by the detailed penis carved onto a bench. The *pièce de résistance*, however, is the enormous town house of a Roman Volubilis oil magnate.

The lower part of the house was a slave-powered oil mill, with one granite grinding stone and two presses. Remains from the mats were chucked into a stone furnace to heat water for the baths. The main stone archway that once held splendid doors leads to a vast reception area. Mosaics on the floor depict Orpheus, god of music and the symbol of the house. The dining room was flanked with low benches so guests could sprawl comfortably as they dangled grapes into their mouths. Next to it was a sophisticated aquarium, complete with built-in amphoras where the goldfish could wait while slaves changed the water.

Hamed made sure I saw the vomitorium. It was a simple rock rectangle over which diners leaned against a chest-high wooden beam. "It was much easier than sticking your finger down your throat," he helpfully explained. "The pressure of the beam did all the work. At the end of the evening, a slave would collect the slop for the tanners." The gastric acid and olive oil were excellent for treating leather.

Up above, in the private quarters, were the solarium, frigidarium, and epluvium. That is, the outdoor playrooms. The standard routine was to sit chest-deep in cooling rainwater, with the upper body, smeared liberally with oil, exposed to the sun.

Back then oil was big business. Excavations at Monte Testaccio suggest that Rome imported 87 million amphoras of oil, each holding about 75 liters, over a 270-year period. About 85 percent came from Andalusia; the rest was from North Africa. Servants rubbed the best oil on the bodies of the ruling class. The rest went for food, light, hair conditioning, medicine, and lotion. By the Monte Testaccio calculations, every friend, Roman, and countryman consumed an average of 22.5 kilos per year.

Archaeology was Hamed's passion, but it was not much of a livelihood in Morocco. Handsome, quick-witted, and fluent in several useful languages, he was desperate to go somewhere.

"I tell everyone who comes that I will work for them four years for free, with only one month's pay for me, if they can get me to Europe," he said. "Or I'll marry anyone. If I could find some woman who was handicapped, maybe blind, who needed someone to take care of her." He laughed, but he was serious. "But only if she is European. If she is Moroccan, she has to be gorgeous."

Meantime, Hamed scrapes by in a little village near Volubilis. He works with wood, weaves kilims, farms, and does whatever odd job he can find. Mostly, he grows olives. His old trees are big and bushy, Moroccan-style. "It can take us all day to do one tree," he said. "Once I was picking with my brother and I couldn't even see him. I kept yelling, 'Are you still there?' "

Last season Hamid took three tons of olives to the neighborhood mill. With little help, this was a big job. But he is no purist about hurrying his crop to the press.

"We throw salt over the olives, and they keep just fine," he explained. "For three tons of olives, you use three hundred kilos of salt. We buy it in blocks and chop it in pieces. Just put the olives in a room and toss on the salt. The oil is wonderful. The best."

Yes, Hamed acknowledged, it gets a little hot down toward the bottom of the pile. Still, that is no problem. And, he assured me, the oil has no salty taste. He had none handy, so I had to take him at his word.

Before going to Morocco, I had been advised to try the olives of Ouezzane. When I mentioned this to Moroccans, each, without fail, nodded in agreement. The small crossroads city in the Rif foothills was in the center of old groves, and tribes in the area were reputed to do wonders with herbs and spices. As luck would have it, the road to Fez took me there.

"Find the pasha and ask him where to go," my informant had said. I found the pasha. His secretary chuckled when I explained what I wanted and why I had shown up with no appointment. She shrugged, picked up the phone, and rattled something in Arabic. I picked up sahafi and zitoun—reporter and olive—which was apparently enough to pique interest in the inner office.

"Pasha" suggested to me a headdress and perhaps even pointy slippers, but these days the word merely means prefect. The man in charge of Ouezzane wore a severe suit, a narrow tie, and steel-rimmed glasses. I repeated my purpose.

"Do you have authorization for this investigation?" he asked.

"It's only olives," I replied.

After a few minutes it was clear that either the Ouezzane region had nothing remarkable to offer, olivewise, or the pasha was not too well informed. I thanked him and headed to the market. Several local citizens were equally perplexed. Nothing I found on display seemed to surpass run-of-the-mill Moroccan olives. I asked around for a while and then moved on.

In Meknes, Ahmed Oumekloul was helpful. He had worked in Ouez-zane. I should find the village of Asjem, six miles down a back road. It was a world-renowned Jewish pilgrimage site. Jews fleeing Andalusia in the fifteenth century had settled around Ouezzane. Their descendants have gone, but people still come every year to the old Asjem synagogue to visit the tomb of Rabbi Amrane Ben Diwan, an Andalusian miracle worker who died two centuries ago. Also, Ahmed said, the Jews' old olive recipes remain.

A day later I was back in Ouezzane, and I rattled up the road to Asjem. The village was no more than a handful of tumbledown houses along a jeep track, but Zoufiq Chengauoue's tiny general store was open for business. Sorry, he said, no more olives. This was September, almost time to pick and cure a new year's crop, and the old ones had long been eaten or sold. Instead, he knocked on a neighboring door and asked for oil.

After a few minutes a bent old woman with whiskers on her chin handed me a chipped glass bottle full of a dark brown liquid. I held it up to the light and noticed several floating insects that looked suspiciously like olive flies. I pointed this out to her, and she eyed me narrowly. With a disgusted snort, she beetled off to fish out the bugs and pretended to bring back a fresh bottle. The oil remained unopened in the back of my car.

The next night I fell in with the kind family of Driss El Marif in Larache, friends of a friend. They insisted I spend the night. My hostess cooked a meal to remember: *harira* soup, fresh sea bass grilled in olive oil, and pigeon *bisteeya*. This was unexpected, and I was empty-handed. Then again, I remembered, I wasn't. If you want to know how that Ouezzane oil tasted, you'll have to ask the El Marif family. They probably loved it.

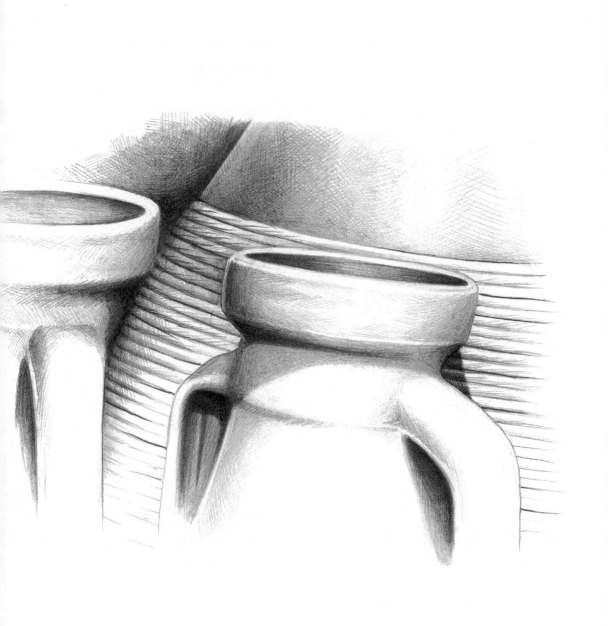

Grecian Oil Jars

Imam Bayaldi

The name means "the imam fainted" in Turkish: a holy man reputedly passed out from pleasure on tasting this sugared eggplant dish, which Greeks now love as their own. Sunny A-Angeles, who exports Elaeon oil with her husband, offers one version of the old recipe.

6	eggplants, medium-sized	3–4	garlic cloves, sliced
2	cups extra-virgin olive oil	2	teaspoons sugar
4–5	onions, sliced into rings		Salt and pepper to taste
2–3	ripe tomatoes, skinned and finely chopped	2	tomatoes, sliced
3	tablespoons finely chopped parsley		

Cut the eggplants in half lengthwise, and soak them in salted water for an hour to remove the bitterness. Drain and fry them briefly in lots of oil at medium heat. Arrange in a baking dish. With a teaspoon, remove half the pulp and keep it in a bowl. In a saucepan, sauté the onions and garlic in oil. Add the chopped tomatoes, eggplant pulp, parsley, sugar, salt and pepper, and cook for about 20 minutes on medium heat. Stuff the eggplant shells with the mixture, lay the tomato slices on top, drizzle generously with oil, and bake in a medium oven until the eggplants are cooked, about 1 hour.

*"A glorious tree flourishes in our Dorian land:
Our sweet, silvered wet nurse, the olive. Self-born
and immortal, unafraid of foes, her ageless strength
defies knaves young and old, for Zeus and Athena guard
her with sleepless eyes."*
—*Sophocles*, Oedipus at Colonus

11

The Lesbos Groves

Olive trees, often, take on the character of the places they inhabit. French varieties are plucked and pampered, *comme il faut*. Tuscan trees are pruned to classic shapes; nearer Naples, they bulge out in generous proportions. In Gaza, limbs sprawl every which way, dusty, unruly, neglected. By the Sahara, they grow tall and noble, like the palms, rising on deep roots. On small Greek islands swept by *meltemi* winds, bent little trunks cling for dear life to barren rock. But on Lesbos, in the haunting pale of dusk, trees loom like great mythical creatures. Ageless, they are living links to the first days of Mediterranean civilization.

Lesbos was an ancient seat of noble thought, scientific inquiry, literary flight, and indulgences of the flesh. Olives figured in all these pursuits. The island's rich groves have expanded and receded with the times, but today its 11 million trees make up the largest concentration of trees in Greece. They are thickest around the Bay of Yeras, a large inlet known locally as the Olive Sea, on which oil-laden tankers ride low in the water from early winter to summer. More dramatic, though, are the monsters and minotaurs that flank the east coast road, from Mitilini, the capital, up to Skala Sikaminias, the tiny fishing port where novelist Stratis Myrivilis landed his *Mermaid Madonna*.

I headed north from Mitilini late on an August afternoon. To the right, sunbeams glistened on turquoise water, which lapped at black pebble shores. Whitewashed houses, ablaze in the hues of bougainvillea, stepped down to the beach. To the left, crowding the narrow road, were olive groves such as I had never seen. In the failing light, glancing rays shone silver far up in the top branches. Centuries of growers had let nature take its course, and the old trees rose so high that pickers could barely reach the uppermost olives with their long, supple poles.

Fat main limbs soared upward from short barrel trunks. Fruiting boughs spread outward, in massive spans. On some trees, thick, bare necks stretched skyward, supporting leafy crowns on top. On others, a single dense mass of foliage swelled from the ground up. Near the village of Mistegna, I stopped to photograph a fabulous old beast with a corkscrewed unicorn horn rooted next to a one-armed giant from the pages of Homer.

"What are you doing in my olives?" a voice demanded from the road. Behind a fierce mustache, a man glared out the window of his battered blue flatbed truck, his bushy eyebrows narrowed with suspicion. A beefy teenager sat next to him. The man's name was Vaggelis Miliorelis. Within minutes, he was pouring us glasses of ouzo on the porch of his modern little house. Interest in his trees, he immediately decided, was a noble purpose.

Miliorelis was sixty-four, the grandson of a refugee from Asia Minor, like the characters of Myrivilis. After the bitter war of 1912–13, the Greeks fled Turkey, and many settled on the northern shores of Lesbos, within sight of their ancestral homes in Anatolia.

"In time," Stratis Myrivilis wrote in *The Mermaid Madonna*,

> they also learned to love the sacred olive tree which for the islander is both religion and sustenance. As the bitterness which had poisoned their minds had abated, they watched with respect the peasants laboriously lugging earth on their backs in wicker

baskets up the steep, gravelly mountain slopes so that a new planting might be added to the thirteen million trees already on the island.

Now they were able to understand the horror of the villagers when the refugees felled these venerable trees for winter firewood. As they became better acquainted with the island people, they realized what an impressive monument to patience and labor was this forest of olive trees that extended in a majestic silver tide from one end of the island to the other. Generation after generation had worked this soil and watered it with their tears and sweat. They knelt before it as one kneels in church to speak with God or as a man kneels before a woman to plant seed in her womb.

Harvesting for half the year, Miliorelis had collected eleven tons of olives from his few hundred towering trees. His wife and grandchildren helped, but he could not afford laborers. They began the harvest late in October, spreading black nets under the first trees to catch olives whacked loose with poles, and kept at it until May. Into the spring, overripe olives dropped by themselves. Miliorelis used an ingenious little machine, like a carpet sweeper with wire brushes, to whisk them off the ground. Birds were not a problem. Helicopter spraying for olive fly had taken care of all such pesky wildlife.

The blue truck shuttled regularly to a nearby cooperative mill, and the final product was sold, like most Lesbos oil, in bulk, to other countries. Drought in Andalusia had forced Spanish buyers to join in bidding with the Italians. Even the oil blenders in Spain had run short. With the European Union subsidy thrown in, there was plenty of cash to toss across the backgammon tables. But the next year, Miliorelis knew, was likely to be lean. He loved his trees, but he diversified into animal feed and other pursuits. If the harvest took half the year, little time was left for pruning. Instead, he lopped off the odd branch to help in picking. Damage from the harvesting poles slowed growth a bit. Still, each year the misshapen giants climbed to even more Cyclopean proportions.

Together, we met the mayor of Mistegna, a fussily dressed telephone company employee named Stratis Stavrakis. Men like Miliorelis were vanishing fast, Stavrakis said. In recent years the village population had dropped by half to just a thousand. Young people went to Athens and beyond, looking for a decent wage, if not simple excitement. "The future of olives is not very optimistic around here," the mayor said, shaking his head. "In a few more years, there won't be a harvest."

This had a familiar ring; all across the Mediterranean, olive-country people worried about the future. But there seemed to be plenty of present left. People loved their trees too much to abandon them. Sure enough, after our second farewell ouzo, that old oliveman's gleam crept into Miliorelis's eyes. "Would you like to taste my oil?" he asked. "It is the best in Greece." Without bothering to wait for a reply, he reached under the sink for a corked glass bottle.

I sampled a spoonful. It was flavorful enough, but it burned like battery acid. I could imagine those bruised and fly-fouled olives lying for days on end until the overworked farmer got them to the miller and his warm-water centrifugal press. "Three, maybe 4 percent acidity," Miliorelis told me, smacking his lips. "It has real body, doesn't it?"

At a taverna table in Skala, under Myrivilis's ancient mulberry, I sat in the past and contemplated the future. The scene was thrilling. A worn stone quay was lined with wooden fishing skiffs in bright reds, yellows, and blues. Drying octopus hung on poles, next to nets in need of mending. Wraiths in black floated over the cobblestones, clawlike hands clutching at headscarves. Across the tiny harbor, the gleaming white Mermaid Madonna's chapel stood high on its imposing rock. Grizzled men with leather faces sat on rickety café chairs and earnestly knocked back retsina.

Down below, in simple houses by the water, were the fishermen. Up above, in a haphazardly laid-out village of great beauty, the goatherders. In between, rising with majesty in terraces up the steep mountainside, were olive trees. Fish, feta, and oil had sustained family after family here for thousands of years.

But above the chink-chank of syrtaki music was the scream of Kawasakis and the blast of U2. This was nearly the twenty-first century. Greeks, like the Dutch or Danes, belong to the European Union, a turbocharged economic bloc that rivals the United States and Japan. The farthest-flung corner of Greece is wired into a new world. A round trip from Athens to New York takes no more time than the ferry from Mitilini to Athens. Worn-out trouser knees and goat manure under the fingernails are no match for Armani shades. Old ways are going fast.

Even more, the world keeps showing up for a visit. For much of each year people from around the globe jet in for a suntan and a *salata horiatiki*. Hardly any Greek island escapes the tourist invasion, but Lesbos has its particular lure.

Nearly two centuries ago, Lord Byron wrote: "The isles of Greece, the isles of Greece! Where burning Sappho loved and sung." Sappho, antiquity's most celebrated poetess, won such praise for her talents with pen and lyre that Plato called her the Tenth Muse. She taught young girls the womanly disciplines of Aphrodite, song, dance, and *delicatesse*. Over two millennia, what else she may have taught has grown to be one of the great mysteries of erotica.

Whether or not Sappho was the first lesbian, the beach at Eressou, near where her academy once stood, is a popular pilgrimage site for gay women couples.

With so many visitors and distant attractions for young Greeks, the people of Lesbos say something will have to change. The olive groves, for instance.

"You can see it happening," remarked Maria Peterson, a German Grecophile I met who had bought a house on Lesbos a decade earlier. "The old men will tell you that their olives are not what they used to be. It is too much work. When I first got here, I used to watch the women come in every night from the harvest. It seemed romantic, but the reality was different. They were bent over in pain, too tired to move. They'd spend hours picking splinters from their fingers, from scooping up the fallen olives. Then at dawn they would go out again. For this, they earned two thousand drachma a day. Now they earn that much by renting a room in their house."

Spare rooms are hardly lacking. Sons are elsewhere, waiting on tables, renting cars, teaching windsurfing, or cruising the beaches for the daughter of a rich foreigner. This was the phenomenon I had seen in Jerba taken to the nth degree. On a small island, where families are large, a few months in the trees can be a handy source of additional income. But the groves in

Lesbos are too big for that. The island is the third largest in Greece, nearly two hours' drive from one end to the other. Some trees climb mountains that are too steep for tractors, far from the nearest vehicle track. The dreaded dacus—the olive fly—is a constant threat. Growing olives is a full-time job.

Michael Binos knows the problem as well as anyone. His 20,000 trees have been in the family for generations. Back during the Turkish occupation, Binos's great-grandfather spent his childhood pushing a plow on his dad's poor patch of land near Pigi, north of the Bay of Yeras. One morning he did something careless and his father slapped him. Without a word, the boy went off to sea. When he came home forty years later, he was the Onassis of his day, a rich owner of sailing ships. He had a new name: Binos. *Bin* means "thousand" in Turkish, and he had made thousands.

Binos bought all the property in sight, olive groves, vineyards, and fields. Soon after his prodigal return, he went to the nearby thermal baths, where he dropped dead of a heart attack. Villagers rallied around his widow, offering to manage her affairs so she could grieve in peace. After a while, Binos's ten-year-old son, Michael, took a close look at the neighbors' kindly assistance. They were selling off bits of his mother's land and pocketing the proceeds. He seized a stick and ran everyone off the property.

Alone, young Michael ran the farm. Later, he added to it. His big stone house at Pigi, still known as the Castle, flew the Greek flag. The Turkish authorities decided not to tangle with him. His own son took over the prosperous olive groves and spent a lifetime making good oil. When the grandson, also named Michael, eventually inherited the property, he was determined to grow the best olives in the Aegean. Before going to work, he studied agronomy in Florence.

"I came back here from university in 1958, and my neighbors wanted to lynch me," Michael told me. "I cut back my trees, sawing off big, high limbs to allow new growth from the trunk. I tried to explain that this would make the trees healthier, the olives fatter and easier to pick. This was heresy. They told me I was crazy, a destroyer of nature. Now most of them have come around." He was pleased at the thought.

At sixty, Binos is rugged and handsome, a well-heeled businessman in faded Levi's and a "USA" baseball cap, clean-shaven with a dirty T-shirt. He starts work at 5 a.m. and seldom knocks off before 11 p.m. Binos does well with his ready-mix concrete plant, his animal-feed business, and his cattle and pigs. The olives worry him sick.

"The trouble with olives is that you need manpower, and we don't have it," he said. "Young people want to work for the civil service or the police, or anywhere else. My foreman hires whoever he can get, even children, and they damage the trees with their poles. For a year I had some Albanians and my head was a little less confused. They worked hard, and well, for not much money. But the government sent them all home. Greeks, they don't want to do this work."

The 1994–95 crop on Lesbos was the best in twenty years, but Binos nearly did not harvest it. Seasonal pickers, knowing he was desperate, bargained hard. "I offered a crew half the olives, and that still wasn't enough," he said. "Finally, we worked something out."

Binos walked me through his trees, pausing regularly to answer the phone ringing in his hip pocket. After one call he tore off in a small truck to resolve some unexplained crisis in a corner of his empire, returning a half hour later.

We plowed through knee-height clover he had planted to add nitrogen to the soil and feed his cows. He showed me a drip irrigation system, a network of thin black tubes along the ground, that helped a selected olive orchard make it through the parched summers. His well-shaped trees, I noticed, were thick with excess growth inside the main branches.

"For me, the world's greatest invention was the chain saw," Binos said a little sheepishly. "It is all we use for pruning. There is no time, not enough labor for anything else." His technique was brutal but effective. He sawed into a thick aging limb to force new buds to sprout below the cut. When the new wood was a few years old, ready to bear fruit, the old limb was buzz-sawed into wood for charcoal.

Binos starts to pick in October, when the olives are green, using the first fruit for extra-virgin oil. If the year is good, his crews might still be collecting olives in July, picking shriveled black lumps off the nets to be sold for refining.

"The secret is, you have to use every bit of resource you have," he explained. Olive-tree cuttings feed the goats and cattle. The heavier wood is piled with pine needles and smoldered down into bagged charcoal; he produces 150 tons of it a year. Old olives go into industrial oil, with no sentiment wasted. Mash from the mill is recycled as fuel. Nothing is lost.

What is missing, Binos said, is oil he can label and sell as his own. He ground his teeth when asked why producers sell their oil in bulk to the Italians and Spanish rather than packaging their own.

"This drives me crazy," he said. "We had a perfect opportunity this year. The Italians had no oil. We could have sold ours instead. When a customer went into a store and could not find what he expected, he would have tried our oil, and he would have been converted. We make the best stuff in the world. But when I am in the United States, I have to go to Astoria, in Queens, to find Greek oil. Why should this be?"

No innocent, he knows why this should be. Greeks could export more labeled oil if they were willing to push a rock up a hill, like Sisyphus, against the Italians' entrenched market share. Binos was looking into it, and he found it daunting. He needed a label and advertising, a bottling operation, a reliable distributor, and no small measure of good fortune.

"How many more jobs can I take on?" he mused aloud. "One man can't do any more by himself. Already I never get to sleep before midnight, and I'm up by dawn. If I hire someone else, there goes my profit."

By Binos's calculations, the market price of oil barely covers his costs. The cream comes in the form of subsidies from Brussels. These, he said, can keep olive growers above water. But subsidies are complex and badly abused, giving an unfair advantage to a small group of people with dubious scruples. Like his neighbors, he was saved by European Union compensation in 1987 after heavy snows cracked and froze millions of Lesbos trees. "They paid for damaged trees," he said with a snort. "And they paid for a hell of a lot of healthy trees." At other times, when prices and subsidies go off track, there are only bank loans, which in Greece approach 30 percent annual interest.

"It is an impossible life, and you have to love it," he said. On this last subject, Binos was eloquent. "You wake up in the morning and smell the smells," he said, "and you know why you do it. You look at the trees, watch them grow, see them renew when you work with them. It sure beats struggling in some city."

But that's part of the problem. Binos's grandfather felt that way at the age of ten. His father worked the same land and loved it, as he does today. Now, edging into his sixties, Michael Binos has a single son to take over the family trees.

"He's a good son, and I love him," Binos said, once again leaving aside his characteristic grin. "But unfortunately he does not care much about olives. He is a graphic designer in Athens. I am the last one."

That night I sat around a grubby table under the stars in the village of Parakila and listened to a crowd of small-time growers talk about the impending end of their world. Most had a few hundred trees and a tenuous cash flow. Mustaches bristled with rhetoric. Clacking worry beads punctuated anguished sentences. Every face was lined and sun-scorched. As the ouzo flowed, I heard more questions than answers.

Who would pick their olives when they no longer could? How could they escape their dependence on Italian buyers? Why didn't the world know about the surpassing quality of Lesbos oil? Why did subsidies from Brussels seem to favor crooks and big business?

Manuel Athanasselis, nearing sixty, in a threadbare sweater and clapped-out shoes, calculated that a grower needed 1,200 trees to survive on olives alone. Others agreed. That happened to be how many he owned, but he still worked as a mason to make ends meet. "One year, maybe okay," he said. "The next year, terrible. Whether the crop is good or bad, the trees need us."

Those with sturdy wives and sons were the least distressed. A willing family could pick a lot of olives, especially if equipped with a curious-looking Rube Goldberg thrasher that was enjoying what would likely be fleeting popularity.

The gas-engined power picker uses plastic straps to flail the olive-bearing branches. It is essentially a cat-o'-nine-tails, minus some tails, at the end of a long pole that is attached to a squat little motor unit. Carefully directed, it shakes the branches hard enough to dislodge ripe olives without chopping up tender shoots. That is the theory. In Crete, where it was invented, some farmers were already abandoning it, the Parakila growers reported. Not only did the contraption damage the trees but it also bruised the olives, raising acidity.

More than anything, the thrasher was a device of desperation. Perhaps half the young people of Lesbos were abandoning the fields. Thinking of Binos's brief experience, I asked about immigrants, or migrant labor on a seasonal basis. Economic chaos at home had pushed hundreds of thousands of Albanians into Greece, and some of them had working papers. Romanians and other Eastern Europeans needed work badly. And many families of Greek blood were coming home from Russia, the way some of their own forebears had left Asia Minor.

When my question was translated, dark looks were exchanged around the table. "Too lazy," said one; "Too shifty," added another; "Lousy workers," chimed in a third. "Out of the question," said Dimitris Kavouras, president of the local cooperative. "They do not fit in here."

Greek villagers, by and large, like their foreigners to bring in money and go home when their vacation is over. Innovation is a hard sell in the face of tradition. With this lack of manpower and resistance to change, I wondered how they took care of their trees. Binos used a power saw on his vast holdings. Would these small growers do things differently?

"We know we should be thinning out with shears and handsaws, but that takes time," said Christos Krelis, thirty-four, introduced as the village's expert pruner. "Mostly we use the chain saw." Up-close fanciwork was a luxury few could afford.

The only one in the group under fifty, he alone strayed from the general thread. Krelis wanted to know what I had heard about organic olive growing. How could they fight dacus and diseases without costly chemicals that upset the local ecology? The farmers on mainland Greece had reported success, I said, and I was on my way to investigate. Also, he was eager to know about the Italian machines that made oil with recycled water at low temperatures. Quality, he felt, was a problem on Lesbos.

There were other worries. Mysterious fires destroyed huge groves on Lesbos. Some growers blamed the Turks, accusing their eternal foes of sneaking across the narrow strip of water to sabotage their main industry. Others suspected careless tourists, who flipped cigarettes out the window of their rented cars. The third theory most likely explained the majority of fires. Developers set the trees alight to clear land for hotels and holiday homes. By the old way of calculation, land on Lesbos was measured by its ability to produce olives. Near the water and in popular tourism areas, values are now figured at so many drachma per square meter. That can be a lot of drachma.

They worried, too, about the rain. In the old days, growers could count on a good winter soaking and then regular showers during the crucial growing period early in summer. Now bizarre cloud patterns formed in August, but the island often went much of the year without rain.

Then there were those subsidies. For a while things were looking up. The authorities had cracked down on thieving millers who fouled the system. Until 1994 subsidies were paid mostly to the people who pressed the oil rather than to those who grew the olives. This meant major fraud. Millers declared far more oil than they actually turned out, and many lied about its quality. New rules changed that. The number of olive mills in Greece plummeted overnight. Farmers earned more, and the oil got better. But Greek officials decided to recuperate the money paid to fraudulent millers. Since most of these had melted into the landscape, the government ordered the growers to refund legitimate subsidies paid to them. This made no sense, of course, because most growers had nothing to do with mill operations; they were merely the customers. It was, as an Athens economic newsletter put it, like beating the saddle because the donkey was out of reach.

By then, the crowd had grown larger, the talk louder, and the worry beads clacked harder. When I pressed the Parakila growers about why they did not market oil under their own label, Kavouras, the coop president, grumbled something harsh to my interpreter. The farmers' job is to produce oil, not sell it, he had said. They all belonged to an island-wide cooperative in Mitilini, which was charged with doing exactly that. He did not seem pleased with the results to date.

Fresh red mullet in oil was waiting down the road. Getting up to go, I asked one last question. If growing olives was so hard and such a financially shaky prospect, would they consider doing anything else? A thunderous Greek chorus answered as one man: "No!"

Athanasselis stood up to make his point. "I have been tending my trees for forty years, my father's trees," he said. "It is what I know, what I love." Laying bricks was to round out the month. But growing olives was a life.

The next morning, Kavouras showed me the village mill, a hangar, and a small tank farm behind his general store. The continuous system was a late-model Alfa Laval that had cost them $200,000. Most people preferred the taste of oil from their junked traditional-style press, he said, but progress could not be ignored. The new centrifugal machinery could handle thirty tons of olives in an eighteen-hour day. It ran until late April.

The mill was not excessively clean, and I suspected the water it used climbed well above 30 degrees. Kavouras had no oil around for me to sample, and I did not particularly regret it. Still, my curiosity was working overtime. What did Lesbos oil taste like?

If history is any guide, Lesbos oil should be wonderful stuff. Growers have had plenty of practice. Three centuries after Sappho, the great thinker Theophrastus wrote his botanical treatises on the island; he was among the olive's early champions. Epicurus devised his guidelines for a rich and reasonable life on Lesbos. His philosophy was not the blind hedonism ascribed to him later by the Romans, but he loved Lesbos olive oil.

From antiquity, Lesbos wine was as good as its oil. "There do not hang from our vineyards grapes such as those harvested in Lesbos from the vines of Methymna," Virgil wrote. But those vineyards were eventually replaced with olive groves, as were most of the island's tall pine forests. By the beginning of the twentieth century, two hundred specially built vessels from the shipyards at Plomari, known as *kaiques*, carried oil to ports from Russia to the far Mediterranean. With planting, harvesting, pressing, and shipping, olives were the life of Lesbos.

Olives, in fact, were the social and economic liberation of Lesbos. Most of Greece ran off the Ottoman rulers in 1830 after a bitter war of independence, but the Turks stayed in Mitilini until 1912. Oil money helped to fund an uprising. Turkish tax collectors had kept families poor by taking 12.5 percent of the crop. But the farmers learned to outwit them by placing flat stones in their piles of olives. Inspectors poked a stick to measure the depth and hit only a false bottom.

With combined resources, small producers built the Aghia Paraskevi cooperative mill in 1910, the first of what became Lesbos's unique system of community presses, which cut production costs and provided money for schools, waterworks, and streets. This broke the monopoly of private mills owned by members of the Turkish-backed elite. Fearing attacks from private millers, the farmers built their cooperative's roof of fireproof metal.

Thanks to the torches of invading Turks over the centuries, the island's trees were periodically renewed. The soil is good: rocky, nitrogen-laced, and well drained. Until recent years, rainfall had been steady, at a generous 700 millimeters a year. Freezes are rare. The staple tree is the kolovi, a vigorous cultivar with small olives, rich in oil.

These days Lesbos produces about 25,000 tons of oil a year, ten times as much as all of France. But local chauvinism aside, Lesbos oil is not known as an elixir of the gods. At the Ministry of Agriculture office in Mitilini, a jolly agronomist from Corinth told me that after seventeen years on the island she still has her family send oil from home.

I asked Alekos Giatzitzoglou, head of the cooperative the Parakila villagers talked about, what was going wrong. He was a friendly man, with a bushy head of curls, eager to help. But he was not optimistic. After the bumper crop in 1994–95, he said, the cooperative had shipped three-quarters of its oil in bulk to Spain and Italy. The rest was packaged as Lesel, with an ugly label, and sold in markets not yet saturated by others. Selling to Americans, he said, seemed to be an impossible dream.

"We must have a better national policy for marketing Greek oil," he said. "Not enough is being done. We need a bigger budget for promotion, advertising. People don't know what we have to offer. It is very expensive to publicize."

I asked him to tell me about his oil. It was good, he said. How good? Very good, he specified. Why should someone buy Lesbos oil as opposed to other Greek oils? He mentioned oil from some islands which, he said, was blended in a dubious manner. But why was his superior to others which were pure? At that, he sent an aide to find some brochures in English, German, and Greek. They said, essentially, that Lesbos oil was very good.

"Look," I suggested, "pretend that I'm an important buyer from the United States. What would you tell me to convince me to try your product?"

He smiled and shrugged.

"Could I taste some?" I asked.

Moments later his aide returned with a quarter-liter bottle and opened it with a flourish. It was rancid.

"Yes, well, this is from the year before," Giatzitzoglou said. I looked at the label, which confirmed that. "We don't have any of this year's oil. If you come back tomorrow, I'll have some." I returned at the appointed hour to find the office had closed early.

By then I had tried hard to find a sample of Lesbos extra-virgin oil. It is not easy. The Greeks consume four times as much olive oil as anyone else. They average twenty liters a year for every man, grandmother, and infant. Americans might like those tiny bottles, but the most popular size container in Greece is a 17-kilo tin. On Lesbos, people visit their favorite mill and buy a year's supply in a barrel. No one, apparently, had thought of packaging oil for the tourist trade.

I persevered and found Stipsi. A small mountain redoubt in northern Lesbos, Stipsi has a cooperative that sells extra-virgin oil under its own label. It was, I was told, the pride of Lesbos. The cooperative was closed when I got there. A local grocer had no oil, but he knew where he could get some if I had a container. While he vanished down a back alley, I bought a liter of mineral water and emptied it into the street.

The grocer returned with a triumphant smile. He produced a wide-mouthed jar, which looked suspiciously like a hospital specimen. We tore up some bread and invited local opinion. Yes, the verdict assured, this was good Lesbos oil. It was pleasant enough, with a clear golden color, a light buttery consistency, and a faint olive taste. Nothing, however, to rave about.

It was at the airport that I finally chanced upon the holy grail. Every type of Lesel oil was lined up on display, in all qualities and sizes, each with their unappealing but brightly colored label. But they were locked in a glass cabinet mounted on the wall, for display purposes only.

Kalamata Olives

Eliopitta (Classic Cypriot Olive Bread)

Diane Kochilas, an American living in Athens and author of *The Food and Wine of Greece*, loves Kalamata olives and oil from the wild Mani. This recipe yields one 10-inch round loaf.

3½ to 4 cups bread flour
1 tablespoon baking powder
   Salt, to taste
¼ cup olive oil

1 tablespoon dried mint
1 cup finely chopped onion
2 cups pitted Amfissa olives

Preheat the oven to 375 degrees. Lightly oil a 10 × 2-inch round baking pan. In a large bowl sift together 3⅓ cups of flour, the baking powder, and the salt. Make a well in the center. Add ¾ cup of warm water and the oil, mint, onions, and olives. Stir with a wooden spoon until the dough is formed. Turn the dough out onto a lightly floured surface (using the remaining flour in small increments) and knead until silky smooth to the touch, about 10 minutes, adding more flour if necessary. Cover and let rest for 10 minutes.

Shape the dough into a ball and pat it down gently to fit evenly in the baking pan. Bake for 50–60 minutes, or until the surface is golden and the bottom sounds hollow when tapped. Remove the bread from the pan and cool it on a rack.

Adapted from *The Food and Wine of Greece*

"The olive fruit is the greatest cure for
any problem of life."
—Solon, who decreed the death
penalty for murdering a tree

12

The Big Kalamata

"Taste these!" ordered Lefteris Papanikolaou, surgeon, politician, and olive chauvinist of the best sort. He held out a clay dish of slender purple-black olives, long, graceful ovals that came to a point, like almonds. It took no effort to comply; I'd know that shape anywhere. I nibbled the first with some decorum and then wolfed down a dozen more at such speed that my host tactfully moved his suture-tying fingers out of the way.

These were Kalamatas, from a tree with the same name as the large port in western Greece just down the road from Lefteris's village of Arfara. They were prepared by a neighbor, payment for some medical service in a place where good olives mean more than money. We ate them in Lefteris's back yard, in the shade of a Kalamata tree, which is as distinctive as the fruit it produces. Its bark is smooth, its branches are long and elegant, and its leaves are big enough to be eucalyptus.

If I had left Lesbos a little disappointed, I arrived in Kalamata country with a happier heart. These salty-sweet delicacies, cured in wine vinegar, were the height of what the Greeks had achieved with five millennia of worshipping the olive. And that was before I tasted the oil.

A different patient had paid with some of his own private oil stock. Lefteris's wife, Vasso, brought it to the patio table in a large bowl, dripping from tomatoes, cucumbers, and feta cheese and collecting at the bottom in an inch-deep pool. This was the best I'd ever tasted. Period. It was hand-pressed and, eight months into the year, still cloudy. With the first taste, I finally realized what oilphiles mean by "peppery"; it had none of the Tuscan bitterness of chlorophyll and rampant polyphenols but rather a wonderful bite that called the taste buds to attention. Quickly it mellowed to a simple, clean flavor of perfectly ripe olives with a nutlike nuance. It was thick but not greasy, with a strong, pleasant aroma and a smooth finish.

That night, at a taverna by the sea, it got better. Lefteris sneaked away to warn the owner to use his best stuff. We ate for hours: sautéed octopus, crisp sardines, fat Mediterranean white fish, calamari, fried eggplant slices, steamed spinach, Greek salad. Everything but the watermelon and half-solid coffee was prepared in the same green-gold Kalamata oil.

Breezes swept the terrace, scenting the night with perfumes from the hills. It was late, but whole families crowded the neighboring tables. Aging grannies in black and young kids in Bart Simpson T-shirts put away huge quantities of food. Their conversation was sometimes slow and thoughtful, variations on ancient philosophical themes. Periodically, it erupted in rumbling laughter.

I had fallen into the heart of the olive's natural habitat, and Lefteris, the doctor, offered his point of view. The health miracles attributed to olive oil were part of a larger framework. A New Yorker who blows gaskets from dawn to *Nightline* but splashes oil on his spinach at lunch will not necessarily see his hundredth birthday. It helps if you can spend the day following your donkey up a rocky mountainside and then sit inert for hours behind a glass of ouzo that goes through your arteries like Drano down a sink. Or, at the least, relax at a taverna.

My status had shifted subtly from honored guest to member of Lefteris's family, and I was having a very good time. When serious, my host looked like a TV-commercial doctor in gold-rimmed glasses recommending a good painkiller. When not, which was more often, he was a sporty-looking favorite uncle who shook when he laughed. His father had been a Greek Orthodox priest in Arfara, living in that same house with the Kalamata out back. Lefteris decided to be a country doctor.

Beyond his generous hospitality and natural warmth, Lefteris had an ulterior motive. He had represented the city of Kalamata and the surrounding Messenia district in parliament, and he planned to run again. Olives were his constituents' only wealth; he wanted foreigners to know about them. Connoisseurs would discover superior quality and shower wealth upon his struggling region. In the olive world, of course, nothing is that simple.

For several reasons, it is hard to find good Kalamata oil outside the region, even in other parts of Greece. The biggest customers are Italian bulk buyers, such as the Fontana brothers of Lucca, who make Filippo Berio. They fly in for the first pressing and bargain hard with a half dozen middlemen exporters. Then containers of fine oil steam off to Italy to improve the quality of Spanish and Tunisian oils for packing under the comforting label *Prodotto Italia*. This was a familiar story by now, but suddenly it had all the pathos of an ancient Greek tragedy.

There was that general marketing problem I had heard about in Mitilini. In the United States all Greek oils combined make up barely 3 percent of the market. The best stuff is seldom available at any price. Yet in Kalamata the problem dates back centuries. In the 1700s, after the Turks kicked out the Venetians, the region's best oil was hauled off to soapmakers in Marseille and Smyrna. When the 1789 revolution crippled the French merchant fleet, Turkish-Greek shippers used Kalamata oil as a cash float to expand their reach. They bought wheat with it in the Crimea and traded that for manufactured goods in France.

Over later years, Kalamata oil never found the world reputation it deserved. But when I sat down on a Sunday morning in Arfara for another olive growers' roundtable, pride was not a problem. The mood was decidedly different than among the villagers in Parakila.

A few years earlier, the Arfara villagers had pooled their resources and brought themselves into the modern age. Their little centrifugal system handles all their olives, without much fuss, recycling the vegetable water and operating at low heat. A balding old farmer named Dimitri Bouras wanted to junk the newfangled thing and go back to the old mats. But the others shouted him down.

These men took great pride in trees they tended with care. If the Kalamatas could be protected from olive fly and drought, they produced tons of those singular olives for the table. At the worst, undersized Kalamatas could be pressed along with the little round olive that made up most of the oil, the koroneiki.

By August, they had already sold 95 percent of their crop at a respectable price. Oil brought them an average of three dollars a kilo that season, and it was heading to four dollars. The middlemen made their profit from EU export aid, nearly a dollar a kilo. It bothered the farmers that their oil disappeared into ignominious blends, but each looked as if he had had a good night's sleep. The main thing was that they had plenty of wonderful oil for family and friends, with enough savings put away to cushion them against a bad season.

As it happens, these villagers were about to be even happier than they imagined. Up their winding road in rural Greece they pay scant attention to the world beyond. Each knows, however, what it means if the rain in Spain does not fall mainly on the plain. By the end of 1995, Jaén producers were desperately short of oil, but there was plenty in Kalamata. Instead of four dollars a kilo, growers would be getting eight.

For the reputation of the table olive, the name Kalamata ought to be golden. Among Greeks, Kalamata oil is widely regarded as the best there is. Greek families living abroad have friends ship it to them in bulk. But in the splendid disorganization of modern Greece, there is not much in a name.

In the early 1990s the parliament passed a law protecting about three dozen regions and towns along the lines of the French *appellation d'origine contrôlée* for wines. It was a noble idea. Qualifying packers printed the origin name in big letters on the label, and their own brand followed in much smaller type. With this safeguard, consumers could be sure of buying top-quality oils from specific regions that had distinct characteristics. They could trust a label reading Kalamata, or one promising excellent oil from Hania in western Crete, or others. Local producers in each area would be spurred to keep their standards high. Their name, and their livelihoods, would be at stake.

But the Greek law defined too many origins to regulate effectively. Spain had just four such designations. France protected only Nyons. In Italy, Tuscany was still angling for its own label. While other countries imposed rigorous enforcement, Greece did little to give its law meaning. Many producers who did not qualify fudged the limits, and consumers had little faith. As a result, few packers took the trouble to comply with costly labeling and inspection technicalities. "Kalamata" on the label is hardly a guarantee.

For one thing, the name itself is misleading. Kalamata olives are grown for eating, and only the small or misshapen fruit go into oil. Kalamata variety trees are found on much of the Peloponnesian peninsula, to the north around Delphi, and anywhere else people plant them. The oil known as Kalamata comes mainly from koroneiki trees, which extend down into the wild region of Mani, a peninsula inhabited by tough-minded villagers with a history of building high stone towers and making Turks miserable. All of this leaves plenty of room for fuzziness in labeling.

Lefteris took me around to meet oil people in the city of Kalamata, a seaport that lost much of its old character to an earthquake. Though lively and friendly, it is weighted down with bleak concrete.

The Kalamata middlemen are a mixed crew, hard bargainers who make their living on a slim margin. They are businessmen, not farmers. Whatever the vicissitudes of nature, the middlemen survive. If the crop is light, somewhere else is probably worse off. That simply means fewer tons sold at a higher price. In 1994–95, for instance, Kalamata growers had a mediocre crop. But the Italians had almost no export surplus at all, and they were desperate.

The few exporters who package Kalamata oil under its own name have made peace with their lot in life. If they cannot match the Italians at marketing, they at least live well in a cozy niche, helped by European Union export subsidies. And they also sell olives that no other region can offer.

Aristedes Dragonas's family has been exporting Calamata brand oil to America for generations. (In transliterating Greek, it is K or C; take your pick.) Its acidity is around .6 percent, but because of technicalities about potential acidity, it can only be called virgin. And it comes in a can, not fancy enough for status-conscious shoppers who want to see the oil behind a classy label.

Dragonas gamely defended his cans. I asked why he did not try bottles. Lefteris picked up the theme: surely a good-looking glass presentation would do the region proud. When Dragonas's occasional partner joined in, he heaved a sigh. "All right, I agree," he said. "I've got a stock of bottles all ready to go, but my distributor insists that he wants cans. Maybe they think it sells better. Who knows? That's what they want, so that's what we send them."

He did well enough in the Middle East and elsewhere. If the Americans insisted on down-marketing his little-known product, it was out of his hands.

In fact, Calamata brand oil was not that same terrific stuff I had tasted in Lefteris's back yard. That is another problem in the region. Olive variety has a lot to do with the taste of oil, but that is only part of it. The best olives go off if they are handled badly or poorly pressed. Koroneikis are grown mostly in small groves, and they are pressed in a number of small mills. Oil packaged under a single label comes from a number of different people. The buyers must know each village and bargain hard for the highest quality. With so much competition, this is not easy.

Below Kalamata, a narrow coastal road skirts wide beaches and snakes between low mountains of high drama. On some stretches, the landscape is stark and beautiful, with wind-shaped olive trees anchored in rock and imposing stone hamlets in the distance. Side trails lead to hidden valleys, lush as the Tuscan hills. Villages are fortified and flowered, each with twisting lanes of cobblestone that lead to café terraces set under flowering vines.

At the bottom, a finger of land known as the Deep Mani is one of Greece's better-kept secrets. Travelers can find a deserted beach in August, hike among the ruins of Turkish outposts, or sip wine in the shade of ancient trees.

Just south of Kalamata, on the way to Avia, a waterside village that Homer called Iris in the *Iliad*, I found an oil middleman from a different mold. Sotiris Plemmenos was an engineer and an Oxford graduate. He had abandoned a lucrative consultancy in England to come back to his olives. He greeted me on his hillside terrace in late-summer uniform, a swim suit and no shoes. Beefy, with a black mustache and a tough guy's battered face, he had soft eyes and a deep, easy smile. His close friend Lefteris had sent me to him.

Sotiris was distressed by all the fraud, inefficiencies, and corner-cutting which blunted the potential of what he knew was the world's best oil. On this subject alone, we talked for hours, until the sun turned red and dropped into the Mediterranean. The main problem, he said, was European subsidies. They enabled crooks to manipulate markets and adulterate oil, crippling people who tried to produce high quality according to the rules.

"This business is full of cowboys, guys who make money the easy way," he said. Philosophical, he figured that getting angry at cheating in the olive trade was like howling at the weather. "They find so many ways to cheat, you can't stop them. The European Union knows, the government knows, U.S. Customs knows, but it still goes on."

By operating within the law, Sotiris said, an exporter's profit margin is often in the range of .5 percent. Unfair competition can drive an honest man out of business. As in Italy, some sellers cut extra virgin with refined or seed oil. But there is a Greek specialty, so common that oilmen have coined a name for it: *aerigides*. "That," he explained, "means a guy who sells air."

The Kalamata cops are still looking for a man they know as Chronopoulos, whom they will never find because he doesn't exist. Some years back, a mysterious *aerigides*—an air jockey—set up shop under the name of Chronopoulos and began, so to speak, to export oil. He reported transactions, taking names from the phone book, and submitted claims for EU subsidies. Over five or six years he amassed huge sums without ever handling a drop of oil.

The man finally tripped himself up with excessive greed. He neglected to pay the Greek value-added tax, a worse failing than merely cheating the European Union. Some bureaucrat eventually noticed all the untaxed sales. The police dropped around for a visit: the business address did not exist, and neither did anyone named Chronopoulos.

Sotiris frowned when I asked about the Italians. His best customers come from Italy, but he hates the thought of Kalamata oil disappearing down a black hole. "The Italians have no choice," he said. "Producers eat their best oil, or they sell it to tourists at huge prices. Land is expensive there, and they can't plant more trees. They can't possibly make all the high-quality oil they need. How else can they get it?"

He blamed the Mafia for driving up costs with massive fraud. "It can be the simplest things," he explained. "A guy goes to a farmer and tells him that he is producing ten tons. The farmer protests that he is getting only one ton. No, the guy says, just report ten tons and give me the papers. That way, the farmer and crook both make money." Not that the Greeks are innocent, either, he added.

The high cost of olive oil, coupled with intense publicity, is causing many Greek families to switch to seed oil. "I can go to the most remote, isolated village in the Mani, where there is no electricity, no phones, and shopkeepers demand that I supply them with seed oil," he said. "This is a tragedy."

Rapidly, Sotiris said, continuous system centrifugal presses are replacing the old traditional mills. Though a shame, this was an economic necessity. There is the obvious cost factor. But modern mills were the only way to protect production from endemic sloppiness.

"The old mills make better oil," he said, "but they are too hard to clean. Just spill some oil on your pants and try to get it out. Let's say a guy comes in with twenty-day-old olives. It takes a whole day to get the smell out because of this one stupid guy. This happens. Farmers take a long time to harvest, to bring in the olives. Maybe it's a sunny day, and someone comes along and says, 'Let's go get an ouzo.' And all your oil is ruined because of the rottenness of one farmer."

With all the problems, Sotiris is occasionally tempted to go back into engineering. Then again, his little patch of Greece is beautiful. When not forced to reflect on business, he is a happy man. For much of the year, he finishes work with a splash off the beach. Then his friends gather around an outdoor table to laugh past midnight and stuff themselves on grilled fish rubbed in what may be the world's best oil.

Besides, he loves olives. An old grove grows outside his door. "These trees . . ." he mused. "Each one is like an old man who has been around for—who knows?—maybe four thousand years. He has seen so many things, so many generations. You cut him down, you try to kill him, he won't die. You just look at the olive and it will tell you a story. The old people always sat under the olive. It had just the right amount of light and air. If you fall asleep under a fig tree, you get a headache. Under an olive, you dream."

Sotiris's dream is to squeeze out Kalamata nectar in the old way and sell it to people who can appreciate it. It is an impossible dream, but so tantalizingly close that it makes him wince.

Just down the hill from his terrace, the oldest building in Avia is a rambling stone structure at the edge of the water. The doors are rotting, and the roof is crumbling away. Its only visible purpose is as a backdrop for gaudy posters: *Miss Sex Appeal '93, Miss Top Model '94, Club Verge; Ferry Boat, Kalamata–Crete.* In fact, it is an old mill where Sotiris's grandfather made oil to remember. In its day, olives from Deep Mani were brought by boat to a wharf that has long since washed away. Others came by wheezing trucks and donkey carts. It was also the family home.

Mani custom dictates that houses must be inherited by sons. When the grandfather died, he had only a daughter: Sotiris's mother. Following the custom, he gave the house to his brother's oldest son, a civil servant in Athens, who did not particularly want it.

"I was a kid, and I begged my father to buy it from his brother-in-law," he said. "I pleaded, 'Daddy, please, we have to have the house.' He wouldn't do it. My mother didn't want it. She had spent too much time trying to keep an old stone house clean. Instead, my father said he would build us a brand-new house, and we would love it."

No one thought much of the Kalamata coast then. Modern families wanted linoleum and built-in cabinets, homes on the hill, with gardens and a wide sea view. The uncle's property was nothing prime. Later, when tourists came, waterfront property shot up in value. Bars and shops sprouted up on every available spot. The uncle waited. Then zoning laws were passed. The old mill could be a mill once again. But its other possibilities were limited, and it could not be torn down. Against any eventuality— and despite Sotiris's repeated pleas—the uncle would not sell, and the place deteriorated by the day.

"Pretty soon it will collapse altogether, and it will be too late," Sotiris said with a fatalistic shrug of the shoulders. As we got up to walk to my car, he stole yet another glance down the hill.

Farther along the wild coast and ten miles up a winding road from the glorious beaches of Stoupa, Fritz Bläuel's large unmarked hangar seems out of place by the stone village of Pyrgos. Inside, stainless-steel tanks and digital control panels, all gleamingly clean, are arrayed under a banner which proclaims in German: *The most important oil change you will ever make in your life.*

The proprietor, a young Austrian in homespun with a curly brown mustache and round metal-rimmed glasses, believes in his message. "People think nothing of pouring the most expensive, highest-quality oil into their cars," he said. "But when it comes to their own bodies, what do they eat? Feh. When you think of the benefits, good olive oil is a pretty small investment."

Bläuel came to backwater Greece in the 1970s to be part of nature and live off the fruit of an ancient land. He nearly starved. For several seasons he earned a few drachmas by picking olives. By the time he met a pretty tourist named Burgi, also an Austrian, he had moved up the scale: he poured oil into bottles from a plastic beaker, setting new speed records. He showed Burgi his beaker. She fell in love with him, and olive oil, and they nearly starved together.

Eventually, with some financial backing from home, the young couple built their plant to package organic Kalamata oil. Now they produce several hundred tons a year, not all of it organic. They also pack Kalamata olives in oil and another variety of wonderful greens in oil spiced with lemon and herbs. After a few years, they could not keep up with the demand in Europe for their Mani brand oil. But, pushing their limits, they also launched Greek Gold in the United States.

Fritz runs the front office, large airy rooms with Shambala rubbings on the walls. His duties include helping his preschool daughter with computer games she can't read. He cruises the hinterlands to stay in touch with the farmers who supply him with oil and olives. Burgi runs the packing plant with Teutonic zeal, policing dark corners for dust or a skewed stack of cartons. Somehow, a fly once turned up in a bottle of Mani; woe to the next one which tries to slip past Burgi.

The heart of the operation is a set of files, scrupulously kept, which lists all the farms that supply oil to the Bläuels. Each is pinpointed on a map, with a history of the trees, details of production, and exhaustive notes. When Fritz is satisfied with a farmer, he agrees to buy a certain amount of olives and oil.

"We look for a maximum amount of oil from traditional mills," he said. "We control the bags and the pressing. We take temperatures and inspect the olives. If there is anything sloppy, the oil does not go into our tanks. It's a careful operation. And now farmers are actually beginning to trust this thing. They see that it works, and they get their money."

Marketing in Germany and Austria was easy, Fritz said. Word traveled fast. He was more worried about the United States. He showed me his new label, complete with the emblazoned words *No Cholesterol*. I could not disguise a snicker, and he laughed too. Cholesterol is in animal fat but it does not appear in any vegetable oil. "I know that's stupid, but I have to do it. Everyone else does. Americans would think mine is the only oil with cholesterol."

Burgi oversees the table olives. Farmers bring in their best dark Kalamatas, which are washed and sorted. Each is scored with two razor cuts. The olives are soaked for weeks in brine and red wine vinegar and then drained. Next they are bottled in olive oil, which yields lovely purple ovals with a sharp sweet taste. The green olives, round and large, are also packed in oil, which is flavored with lemon rind and herbs. The liquid is as good as the olives.

The hard part has been making organic oil, a new concept in Greece. Fritz and Burgi went five years before bottling a drop of it. First they had to convince the farmers to shift to biological methods, shunning all chemicals and using only plant-based methods to discourage pests. This requires a three-year transition period. In the first year someone sabotaged it by spraying groves from a helicopter. Then they had to win over recalcitrant neighbors. Organic farming works best over wide areas.

The idea is less to kill insects than to deflect them elsewhere. If an olive trunk is dusty with wood ash, for example, wood borers are likely to wander off to the nearest oak. The bigger the area which pests have been persuaded to leave, the less the chance of a sudden attack. A European Union laboratory at the University of Cardiff in Wales, of all places, has been testing natural ways to repel the worst scourge, the olive fly. The most successful trick so far is to bait them into traps or onto old-fashioned sticky paper.

When I saw the Bläuels in 1995, they were happy with the organic side of the business. They had sold a respectable crop, and their list of suppliers was growing. At the end of the year I phoned to see about the new crop. The dacus had struck—and laughed off the traps hanging in wait. Fritz, short of organic oil, had to rely on conventional oil sales to get through the year.

On my way to Kalamata I had swung north from Attica into the heartland above the Peloponnesian peninsula. I wanted to visit an olive packer named Gatos, and I also wanted to see the trees in his neighborhood. This was Delphi, which the ancients believed to be the center of the earth. Back then, naturally enough, olives were held holy.

Balding, frail-looking, and well past the age most people retire, John Gatos rises early every morning and hurries to his olive factory. He works at a simple metal desk under the sepia-tone gaze of his father, who started the business, Gatolive. Behind his office is a big barnlike plant with a raw concrete floor.

Gatos was late for our meeting, so I wandered out back to see the olives being cured. I saw only a warehouse with a few desultory workers. After a few minutes I realized I was watching the action. Industrial olive packing is the simplest form of cottage industry. The olives are washed and run along a conveyor belt with holes in it to drop them into bins according to size. They are placed in plastic barrels, where they sit until ready—two months maybe, sometimes up to six.

That is the process for amfissas, also called conservolia, fat black olives that resemble the manzanillos of California. Kalamatas are first run through a slicer, a rudimentary machine which cuts nearly to the pit so the brine can penetrate. Vinegar is added to the salty water, one part per thirteen of brine, except for olives bound for Israel. The Israelis don't want vinegar.

The length of curing depends on water content and customer preference. At the end, the white fermentation scum at the top is skimmed off. A layer of clear paraffin oil is poured across the top to seal it, and the barrels are carted to a truck for shipping. Or they are packed into smaller cans or plastic jugs.

This seemed simple enough, until Gatos began to explain the nuances. Like making oil, packing good olives depends on a series of small steps, any of which could ruin the end result. Too much salt affects the taste. Too little allows nasty bacteria to form. Grading is crucial. An extra large is not a mammoth, nor is a giant a jumbo. Super colossals come 110 to a kilo, big enough to eat with a knife and fork. Superiors—that is, tiny ones—work out to 280 per kilo.

Every year fortunes change. When the weather is good and the trees produce, Gatos's amfissas can turn out twelve thousand tons. Sometimes they yield barely three thousand. Even more than the oil makers, olive producers fear the dacus. Ugly fruit can still be pressed, but they cannot be put on the table.

I had already gotten a lesson in the economics from John's brother, Nicolas, at Gatolive's hole-in-the-wall office in Athens. He has been president of the Greek Table Olives Producers' Association since 1970, and no one, it seems, wants him to stop.

In Greece, 65 percent of table olives are black, 20 percent are green, and 15 percent are Kalamata. Most of these are simply packed in brine. That shriveled, marinated "olive à la greque" is a generic category, like French fries, usually found elsewhere. The throumbolea is so sweet that it can be eaten black with curing. Altogether, Greeks export something in the range of fifty thousand tons a year. And every year is a struggle.

The overriding problem, Gatos explained, is that successive governments have yet to balance the conflicting interests that emerged after the tumultuous war that led to independence from Turkey in 1830. Greeks like to say they've been growing olives for four thousand years, which they have. But not without some interruptions.

The Turks burned many olive groves during the 1820s, and for decades landless farmers were reluctant to plant new trees. Successive Greek governments promised to break up the Turkish feudal system of large holdings, but it took a century to do it. In 1855, by one count, Greece was down to 3 million productive trees, compared to perhaps 120 million today. Even now, traces of the old system leave many small growers at the mercy of middlemen, indebted to bankers, and dominated by local elites.

Early governments contracted tax collection to private entrepreneurs, who promised fixed amounts to the state. That way central planners knew how much income to expect. Collectors squeezed out whatever else they could as profit. That created inequalities which still remain. Competing exporters in different parts of Greece operate on a split-level playing field.

Then there is inflation. The National Bank of Greece charges interest on operating capital that might reach 28 or 30 percent. No one can afford to sit on stock or extend the sort of credit that buyers can find in other countries. Producers routinely default on their debts and sink out of sight.

Despite all the help given to oil makers, the European Union has no subsidies for table olives. When oil prices are high, olives are sold to the mills, leaving a shortfall for processing. Greek governments have passed various laws designed to support producers, Gatos said, but most have backfired, causing calamity.

"Most of the olive troubles are created by interference of the government," Nicolas Gatos said. "They want to help the farmers, and they make a mess of it."

I saw his point. At the time, the Agriculture Ministry was trying to recuperate subsidies paid to millers who had lied about how much oil they produced. After the European Union tightened its policies, many closed down and disappeared. So the government asked 40,000 growers who had taken olives to these mills to pay refunds. The farmers, who were only customers, had nothing to do with the scams.

While his brother tears out his hair in Athens, John Gatos pickles away in his plant by the sea and spends long hours among the family trees. On a sunny morning he loaded me into his rattling pickup, and we took a tour. His smile widened with each kilometer we drove in the opposite direction of his overburdened desk. Together we clucked over bug damage and yanked away the odd sucker at the base of his trees. We inspected the small canals that brought in water when the rains failed. For breakfast, he plucked fresh green figs off a dusty tree.

"I planted this sevillano as an experiment," he said, showing me a hardy young tree with olives the size of golf balls. "The problem is that if there is one fly in the orchard, he heads straight for this tree." Looking closer, I saw that most of the firm green olives were pocked with telltale brown spots.

Noticing some other trees, I tried to guess their age. Thirty maybe? Perhaps a few years younger? "Six years old," Gatos said, beaming at my reaction. He had fed them, watered them, and coiffed them like pets. They expressed their gratitude. "I don't think," he concluded, "there is anything else like the olive tree."

From Gatos's seaside village of Itea, I drove north toward the foot of Mount Parnassus to Delphiland. Enough remains from antiquity to suggest the sweep and power of the place. After the first temple to Gaia, the earth mother, was built twenty-seven centuries ago, Delphi was the spiritual headquarters of the Western World. Pilgrims came to hear wisdom from the Oracle, which was delivered by a frenzied priestess and interpreted by a priest who usually spoke in verse. The site should overpower, a shrine city in perfectly hewn stone set against soaring cliffs on a mountain that falls away sharply to the turquoise Mediterranean far below.

Climbing up among the ruins, I tried to take myself back three thousand years. I could handle the omnipresent signs that barked orders in four languages: DO NOT CLIMB ON STONES. DO NOT TOUCH. BE REVERENT. I managed to ignore the people climbing on stones, touching, and being irreverent. But the noise got me. Kids yelling, old people yakking about trivialities at home, Greek swells hitting on the better-looking young tourists. I skipped the theater at the top. From the Temple of Apollo I fell in with the brisk let's-check-off-this-dump-and-get-rolling pace of people headed down the hill.

Halfway along the path, I tried again. By the Sacred Way I slowed down with intent to ponder. Just then I heard a shrill voice wail in outraged French: "*Mais alors*, he's going to stop." To my right a woman whose aspect was less impressive than the surroundings was posing for a picture. To my left, at the other side of the walkway, another woman held a camera.

My pausing to take in Delphi would have made her wait a moment. Or move.

I repaired to an olive tree to ponder undisturbed. The tree had been assaulted with a chain saw years before and abandoned to its lot. Looking around, I saw others like it. This was similar to what had happened in Dougga in Tunisia, but on a grand scale. Wherever a centuries-old olive limb began to poke toward a path or a road, it was whacked off with a chain saw. Amputated trees, grotesquely unbalanced, had compensated for their loss by flinging out thick skirts of shoots from the root. No one had bothered to cut them away.

Elsewhere in the precincts of Delphi, the old olives were in lamentable shape, tangled with undergrowth. Whatever the worn stones revealed of ancient Greece, the trees were a telling monument to modern times— a shift from glory to grubbiness—and only they seemed to bridge the time between past and present.

Pruned and allowed to stand in their proper setting, symbols of everything Greek since Athena, the olives would have brought ancient Delphi to life. Authorities seemed not to see it that way. Tourists did not pay to visit trees. No one was paid to care for them, and they were left to deteriorate along with the dead marble and cut stone. Neglected, the olives hover in the shadows like haunting ghosts, visible only to those who look for them.

Gazing down from Delphi, you see a different picture. Majestic groves rise and fall on undulating hills all the way to the sea below. Light catches their leaves in bright flashes and suffused glows. The guidebook cliché is right on the mark: an ocean of olives.

I looked down at this timeless setting and imagined an unbroken line of Oracles, each offering her wisdom for the age at hand: Invade Carthage. Beware of Turks bearing gifts. Open a restaurant in Queens. Divorce the jerk. It was getting late, and I was getting dopey. But when I returned to my car, I found that Delphi had left me a message. A smudgy Xeroxed flyer on the windshield told me which discotheque I would miss that night.

Back in Athens, Antonia Trichopoulou may be the closest thing modern-day Greece has to an Oracle. A world-class nutritionist, she is a favorite of the international olive set, and her message is simple: Eat oil. She advocated the health benefits of olive oil long before anyone talked of the Mediterranean diet. Since then, she has been filling in the details with medical science.

In her austere office at the Athens School of Public Health, Antonia's fax is generally smoking. Someone is always asking advice, passing on technogossip, or urging her to fly halfway around the world to give a speech. Now gracefully graying, a matron with gold earrings and arresting eyes, she seems equally suited to lecturing professors or dancing on the tables. She is witty, colorful, and, on the subject of olives, a downright zealot.

She publishes often with her husband, Dimitrios Trichopoulos, who went to the Harvard School of Public Health in 1989 to be its chief epidemiologist. They manage to see each other every few weeks. No olive-oil congress is complete without Antonia and Dimitrios holding hands like honeymooners.

As a young doctor in the 1970s, Antonia was puzzled by the prevailing advice that lowering cholesterol automatically meant sharply reducing fat. That might apply in heavily industrialized societies where people thrived on animal grease. But Greeks had low rates of heart disease, and they lived on olive oil. It is monounsaturated and wonderful stuff, but it is still fat.

Back then, on her own television program, which dealt with health matters, she came across a middle-aged museum guard from the Peloponnese. "He was miserable," she remembers. "He said, 'I hate soya and corn oil, but they told me to use it.' Kalamata oil is the best in the world, but his doctors wouldn't let him touch it. That made no sense, and I was sure it was wrong. It has taken a lot of work to get together the medical evidence."

Her studies since then have stressed the fact that while Greeks get up to 44 percent of their calories from fats, they suffer much less coronary disease than societies which rely on animal fats and seed oils. That message is echoed increasingly by specialists in the United States and elsewhere in Europe.

"It's crazy for us to change," Trichopoulou said. "Our oil has proved beneficial to the Greeks for a very long time. And we love it. Our main concern is not to impose this idea on others but that others should not impose on us."

But cholesterol is only part of it. On his side of the ocean, Dimitrios studies olive oil's effect on different cancers. Antonia handed me an article, hot off the fax, which she and her husband, along with other specialists, had just published in the *Journal of the National Cancer Institute*. It concluded: "There is evidence that olive oil consumption may reduce the risk of breast cancer, whereas margarine intake appears to be associated with an elevated risk for the disease."

She said that previous studies in the United States and Canada linked a diet high in saturated fats to ovarian cancer. Her survey looked at 820 woman with breast cancer and 1,548 others who had no cancer. It found that the cancer risk was 25 percent lower among women who consumed olive oil more than once a day, particularly women past menopause.

Normal breast cancer rates are 50 percent lower in Mediterranean countries than in America, according to Antonia. Greek women who eat the least amount of olive oil still consume more than American women who eat the most. Because of this, the Greek doctors said, American women might reduce their breast cancer risk by half if they consumed more olive oil in place of other fats.

When I telephoned Dimitrios at Harvard, he spouted enthusiasm. Cautious as scientists always are, he was nonetheless encouraged by recent experiments with rats and colon cancer. He was also studying olive oil's effects on aging. With his wife, he had just published a paper in the *British Medical Journal* which showed that a sampling of Greeks lived longer if they ate olive oil.

Scientists are just beginning to learn the full value of olive oil, Antonia explained. "There are more than one hundred minor components that have not yet been studied," she said. "I don't think that in the next ten or twenty years we will be in a position to know all the secrets."

It was Antonia who introduced me to Papanikolaou and Bläuel. She had visited Kalamata and came away horrified at how poorly producers exploit their singular resource. "They gave me eggs for breakfast, and I sent them back," she said. "They asked me what I wanted, and I said olives. Kalamata olives." None were on hand. Part of the problem was national bank interest rates. Money was so tight that few producers could afford to cure enough olives for the local market. "When I left, they made a big present for me of their best olive oil." She added, sputtering at the memory, "It was in a plastic bottle. Can you imagine? Plastic."

Within Greece, aggressive seed-oil marketers had made deep inroads, and olive-oil producers, made more heavily dependent on exports, had been eclipsed abroad. "It is a bitter feeling," Antonia said. "We have spent so much time researching the benefits of oil, and now everyone else is taking advantage. Our oil comes in hideous bottles, or in tin or plastic. Or it is sold to others who charge tenfold the price. It is not a very optimistic picture."

Antonia sighed and returned to a happier theme. For Greeks, she said, the benefits of olive oil are part of a holistic approach in a diet rich in vegetables and protein. Pressed olives produce a natural juice, with none of the high-temperature chemistry that goes into seed oil. Despite the cheaper price of refined oil, 70 percent of all Greek oil is still extra virgin.

"It is part of the way we live," Antonia said. "Take wine. Greeks never drink it by itself, never outside of a meal or with company. We don't drink wine to forget life. We drink it to enjoy life."

I made a few stops in Athens to learn more about marketing problems. At Elais, S.A., the biggest olive-oil producer in Greece, I met Gregory Antoniadis, the company's oils manager. His card was embossed with the familiar *U* of Unilever, the parent conglomerate. But Antoniadis was also representing Sevitel, the Greek Producers' Association. A smooth talker in a silk tie, he said yes, there was a problem, and no, there wasn't.

Back in the 1960s, he explained, most Greeks went straight to their favorite farmer and lugged back oil in fifteen-liter cans. A decade ago, the dominant packer was Eleourgiki, a cooperative of 350,000 growers, which not only sold oil but also helped the European Economic Community distribute subsidies.

Today, many more Greeks want a label, and marketing is more sophisticated. Consumers bought 20,000 tons of oil from stores in 1980, and now they are up to 130,000 tons, a 500 percent increase. This is good.

But the internal market is stagnating, and this is bad. Seed oils are cheaper, and heavily advertised. And there is little room for the olive-oil market to expand. "This is not a new product in Greece," Antoniadis said. The answer would be to sell more packaged oil abroad, but it is not working out that way. The Americans are the real problem; they buy 70 percent Italian oil, 16 percent Spanish, and less than 3 percent Greek. The Greeks can't succeed until they succeed.

About fifteen Greek brands reach the United States, but none is predominant, and all are hard to find. "If you haven't got that critical mass of sales, you cannot enter the market," Antoniadis said. "We are not meant to be in a niche," he added. "Our goal is the mass market. Someone has to invest, to prepare the ground. The Spanish spent money marketing their oil, but we have not. The producers can't afford it, and the government doesn't do it."

This made sense. It sounded like a more elegant version of that conversation I'd had with the Lesel cooperative people in Mitilini. He did not offer any fresh oil to try, either.

From Elias, the monster multinational, I went across Piraeus to Elaeon, about as far to the other extreme as you can get and still exist. Sunny A-Angeles and her husband, Polydoros, had started the mom-and-pop Mediterranean Nutrition Ltd. With little help they market a peppery yet buttery oil from organically grown olives in the northern Peloponnese.

Elaeon's clear pint-sized bottle comes sheathed in a colorful cardboard box to keep out the light. Each bottle is accompanied by one of a dozen little flyers that Sunny designed for the purpose. Each has a biblical quote or a snatch of poetry, three recipes, and a section of "Grandmother's Secrets," olive-oil folk remedies for such things as frizzy hair, constipation, insect stings, burns, hair loss, bad nails, scuffed furniture, and too much ouzo. And each has a message for the buyer: "You are evidently a gourmet of discerning taste . . . We wish you *kali orexi*—a good appetite."

It was too early to know if this charming base flattery had made an effect, but Sunny and Polydoros were hoping for the best. They are a friendly pair. Sunny was a teacher, a Japanese-German who came to Greece by way of Australia. Polydoros, as Greek as his name suggests, was a businessman.

Like the Bläuels, they market oil from a group of farmers who have decided to stop using chemicals. Theirs is from Egio, near Patras, just across the water from Gatos's place at Itea. Elaeon sells for about $10 a liter, which is a lot for Greek oil. Their total volume is three to four hundred tons, which they hope to move whether or not the government runs interference for them.

"We are just starting to sell in Germany, but our main target is the United States," Sunny said. With a little laugh, she added, "I think if we had really thought about it, we would not have plunged in."

Exploring the olive groves of Greece can be a lifelong pursuit. The trees are beautiful in Corfu. They are old and dramatic in Crete, particularly near Hania, where the Mediterranean Agronomic Institute has an olive research station. Hania produces one of my favorite oils, a mellow golden extra virgin called Kydonia, with a label showing a serious-looking Cretan in long black mustache and bloomers. But I wound up my trip with a pleasure stop at the Byzantine fortress-city of Monemvasia, at the southeast tip of the Peloponnese.

Unlike most ancient ruins, Monemvasia is full of life. Beautifully restored holiday homes alternate with the crumbling apartments of year-around residents. Athenians come over by fast ferry from Piraeus, but I drove an hour or so from the Mani across the bottom of the peninsula. The city is carved up the seaward side of a small but formidable mountain on an islet linked to the mainland by a causeway. Travelers cannot see it until they are at its arched stone gate. Its walls and ramparts look as if they could withstand nuclear assault. But lacking a water supply or fields to grow food, Monemvasia changed hands whenever an army had the patience to lay siege.

The steep narrow streets were thick with tourists, but that also meant a few excellent restaurants to feed them. I found an outside terrace, brilliant with geraniums and blooming vines, where I ate fresh bass above a sweeping panorama of sharply pitched stone rooftops, ocher tiles, and flowers. Far down below, the Mediterranean sparkled in every shade of green and blue. Haunting music spilled from behind weathered wooden shutters.

Monemvasia was built in the sixth century A.D. on two levels. Commoners lived in the Lower Town, which once numbered almost eight hundred homes and forty churches, all jumbled one atop the other along twisting cobbled lanes that run up and down off a single main street. The only other gate leads to a small beach, now a favorite spot for swimming and clothing-optional sunbathing. People of consequence inhabited the Upper Town, with its own crenelated walls high above on the mountaintop.

Late in the day I explored the lower part of town and then headed to the top. This was a serious hike, a half hour of huffing and puffing up a trail over which countless generations had hauled their groceries. Most of the Upper Town has collapsed, leaving a jumble of hewn rocks that wear away a little more every year. Only Saint Sophia remains, a beautiful structure of carved rock and timbers, richly appointed by the faithful.

In front of it was an enormous olive tree, so wild and dusty that I had to look closely to make sure. A Kalamata. The church was built seven hundred years ago, and the tree might have been planted then. I had never seen a tree in such bad shape.

The names of forgotten souls were carved into the trunk, some so long ago that they are now engraved an inch deep. Thick limbs had been amputated far from any joint, so the cuts did not heal, leaving the wood vulnerable to insects and disease. So many dead branches and twigs fought for space that the tree looked as if it were sprouting brooms. Other branches were simply broken off, and ugly jagged ends poked out from the crown.

Someone had lit a campfire on the table-like surface where a gigantic old trunk had been sawn away, and the wood was still badly charred. Someone else had apparently taken an ax to it. Yet another lunatic had looped wire around the tree, nearly ring-barking it to death. Over time, the gnarled root had thrown out hundreds of suckers, and they all sapped strength from the trunk. A few of them were as fat as a twenty-year-old tree.

It seldom rained atop Monemvasia. Hard-packed dirt and rocks around the base had not been turned in a generation. The soil must have been leached of nutrients years ago. Thousands of feet have scuffed overexposed roots, leaving only bare wood.

I had an irrational urge to spend a few days with a pruning saw and a hoe, maybe hauling some nitrogen-laced fertilizer and buckets of water up that steep trail. But it was late summer, and I could see that the tree didn't need me. Despite it all, the big Kalamata had produced yet another crop of fruit.

Italian Pintarolos

Skampi na Buzaru

Snjezana Vukic, an Associated Press reporter whose skills extend to the kitchen, put on paper a recipe that Dalmatian women just south of Split seem to know from birth. Croatian *skampi*—unlike Italian *scampi*, which are merely shrimp—are sweet langoustines from the salty Adriatic. *Buzaru* just means sauce. Proportions are approximate, since local cooks seldom measure.

1 pound *skampi* (large prawns will suffice)
Olive oil, as needed
2 garlic cloves, sliced

2 tablespoons chopped parsley
Salt and pepper to taste
¼ cup dry white wine

Wash the *skampi* and coat ᵗhem with oil. In a deep skillet, sauté the *skampi* in a thin layer of oil, stirring, until their color shifts. Add the garlic and parsley, with the salt and pepper. Stir together and cook slowly. Large *skampi* can take 10 minutes, but prawns are done when pink. When they are nearly finished, add the wine and a bit of water. As soon as the liquid comes to a boil, remove them from the heat. This dish is usually served only with bread. Eat it with the fingers and use the bread to sop up the extra sauce. It can also be served with *blitva*, a vegetable with no apparent translation because it is found only in Dalmatia. Swiss chard comes close.

Serves 3–4.

*"The time of universal peace is near. Prove
this a prosp'rous day, the three-nook'd world shall
bear the olive freely."*
—William Shakespeare, Antony and Cleopatra

13

The Olive Branch of War

After a trip into Bosnia to cover the war, I drove down the Croatian coast
in search of refugees. I found thousands in the beach town of Makarska,
an erstwhile resort of white buildings and luxuriant foliage. Muslims had
settled into the Riviera Hotel until a mob of Croats, recently evicted from
Western Herzegovina, had run them out to move in themselves.

The seaside hotel was badly run-down but still a dreamlike haven for
families whose homes across the border had collapsed into smoking rubble.
All over the sprawling grounds huge old olives offered beauty and shade,
but not, according to their reputation, peace. But they were a mess, skirted
with tall suckers and dense with unpruned branches. When I asked the
woman in charge who was looking after them, she replied with a loud sigh.

"Those trees . . ." she said. "We have simply abandoned them. There's
nothing else to do. All these refugee children climb them and break the
limbs. The worst part is, they tear off all the olives. They play war, and

the fruit is their ammunition. They throw olives as though they were bullets."

Until fighting began in 1991, much of what used to be Yugoslavia was lovely and placid olive country. Old orchards spread down the Istrian Peninsula near Italy along the Adriatic coast, through Zadar, Split, and Dubrovnik, into the neighboring state of Montenegro.

But along the Dalmatian coast in southern Croatia olives were yet one more casualty of the war. Handsome old trees were reduced to forlorn shapes, glimpsed out the window of cars and buses by people in a hurry. In many cases, families were run off their land by one army or another. Frequent roadblocks discouraged casual trips away from home. Oil mills, if not destroyed, were shut down for lack of parts and personnel. Even in extended periods of calm, no one wanted to harvest olives in the potential range of a bored sniper.

Still, a lot of handsome trees, of old lineage and impressive productivity, survived intact. The finest groves are on the islands, such as Krk, near Pula, or Hvar, off Split. The best of them make the golden oil of Korcula.

Having grown up among Mexican cooks and lived eighteen years in France, I've gotten to be picky about what counts as a memorable dinner. But the tastiest meal I can remember was at a rickety outdoor table at the port of Vela Luka, on the Croatian island of Korcula, a ferry ride from Split, just beyond earshot of the war in Bosnia.

Korcula is one of the splendors of the Mediterranean. Its beauty is low-key, hardly overpowering at first, but revealing itself in stages as your car rounds a bend into a dramatic seascape of white foam crashing high above the deep blue. Houses of old stone and faded red-tile roofs step up the hills. Great sprays of bright flowers tumble over whitewashed walls.

The little restaurant, called Pod Bore (Under the Pines), was an open barn with a sizable bar for townsfolk who had nothing better to do in a country with an economy that had collapsed around its ankles. Its terrace was paved in gravel and crushed shell, a net's throw from the fishing boats of Vela Luka. A photographer friend and I sat down to try our luck. The owner's wife brought us fat, fresh Adriatic crawfish bubbling in oil so rich that dessert was bread soaked in what was left at the bottom of the sizzling platter. When we'd finished wiping the dish to a greasy gleam, we ordered more.

Long before Marco Polo left home on Venetian Korcula to bring back pasta from China, this remarkable olive oil was already in ample supply. Polo fought with the Venetian fleet to defend the island against the Genoese in 1298. He was taken to prison, where a Pisan cellmate took notes on long conversations and ghostwrote his immortal travel book.

Venetians doubtless planted olives on the island; the oil trade was a basis of their economic power. But the island, described at length by Pliny the Elder, was first settled by the Greeks. Vela Luka, at the north end, is smaller than the walled city of Korcula, a medieval jewel at the opposite extreme. Korcula's narrow stone lanes are perfectly placed to watch cooling Adriatic breezes. The peaked tile roofs let in the gentle morning and afternoon light but block out the noonday sun.

These days, Korcula's groves are a mix of knobby old trunks with full crowns and scraggly young plants that clearly need someone's time and energy. They are no particular attraction. The word "olive" does not even appear in the 159-page official guide, which includes sections on vegetation, food, and local wine.

My hasty questions got evasive answers on that first trip, and I did not push. On Korcula, in mid-1993, there were more urgent matters than olives. The International Red Cross had just brought several hundred prisoners from a Croat concentration camp, Dretelj. Young men had lost up to sixty pounds. A week after their negotiated freedom, some were still too weak to stand. We hitched a ride in a small boat to their temporary shelter, an old monastery on an islet in a turquoise sea, near Korcula city.

Ivo the boatman was solemn for part of the way back. Soon, however, his irrepressible humor got the better of him. In odd bits of English, he told me about his funky little craft. "Warm eat, dry sleep," he said, pointing first to a bottled gas hot plate and then to a disheveled mattress in a tight space down below.

Back in port, Ivo led me to a friend who knew about olives. As the director of a tourist agency, the man had a lot of time on his hands. He led us on a quick tour of the town's old stone spires and magnificent medieval gateways. Then we settled at a café table in one of those places so beloved by those who frequent olive trees.

The terrace was shaded by broad grape leaves from gnarled climbing vines, thick as tree trunks. Scented jasmine covered a nearby wall and purple wisteria blooms dropped from overhead. Pleasant aromas wafted from the kitchen. It was late morning, and Ivo went straight for the whiskey. His friend and I knocked back slivovitz, plum brandy, in preparation for serious business: they had to decide where I would buy my Korcula oil.

The debate, at a furious pace, lasted into the lunch hour. My Croatian is far worse than Ivo's English, but I believe I caught the drift. Ivo insisted that one goes to Branko for wine and Ratko for oil. His friend had two different names. Finally, I told them I had to head back to the war. Could they reach a compromise? They did, and a detailed map appeared.

Winding my way through back lanes, I found the tiny grocery shop. The grocer produced two liter-and-a-half plastic bottles full of oil. Their Pepsi-Cola labels were in mint condition. Since I was a foreigner deserving deluxe treatment, he wrapped them in a sheet of newspaper. It was rich, mature oil, a deep gold with a distinct but not overpowering flavor of olives. Like the oil I had tasted in Vela Luka, it was perfect for anything Ivo the fisherman might haul up in his nets.

The war had sent prices soaring. I had to pay, I believe, five dollars a bottle. It was a different marketing strategy than Laudemio's, all right, but the taste was full and fresh, and the price was hard to beat.

At the end of 1995, I was back in Croatia again, right in the middle of pressing season. As before, I was on a tight schedule. A sort of peace had been declared in Bosnia, and American troops were about to land in Tuzla. But the road to eastern Bosnia followed the Dalmatian coast south until it turned inland. On the way, I wanted to interview refugees still on hold at the Riviera Hotel. The Makarska mill was just down the road.

Before setting out, I stopped for supplies at the market under the ancient walls of Split. Under a huge red umbrella near the cabbages, a hefty, balding man with a big black mustache stood behind a table covered with bottles. Most were dark green and unlabeled. Some bore the markings of a nasty local brandy. Others said *Rum*. A few were Campari bottles, but the liquid inside was not red. As I hoped it might be, it was the golden oil of Korcula.

Sime Zvone owned 2,500 trees near Vela Luka. He grew the region's popular oblicas, picked in November and December when they are large and black. Like most Croatian farmers, he was short of labor. He had to borrow a truck. After each day's picking, he submerged his olives in water to protect them from acidifying. Every week or so, he hauled them to the cooperative.

For a change, here was an oil maker who had nothing bad to say about the Italians. Zvone had no dreams of selling estate-bottled oil at Dean & Deluca. He would not have minded selling in bulk had any Italian buyer put the Croatian coast on his itinerary. Instead, he put basketloads of liter bottles on the ferry and made regular four-hour round trips to the Split market.

"I make deals where I can," he said. "It is not easy." Our deal was simple enough. I gave him the equivalent of seven dollars for a bottle of oil that tasted fabulous.

A handful of others sold oil at the market. Some pressed a smaller olive, called lastovke, which produced even better oil. With less water, it yielded a liter of oil for five kilos of fruit, the way our best olives do in France.

Near Zvone, an old woman in a ratty fur hat displayed cured black oblicas in an enormous plastic sack. They had been picked only days before. On the Croatian coast, curing is simple. People dump olives in nearly boiling water and leave them for ten minutes. This zaps the bitter oleuropein and also the flavor compounds. Mushy and tasteless, they are awful.

An hour south of Split, I found the Riviera Hotel in its same sad state. The refugees I had met from Western Herzegovina two years earlier had gone, replaced by new ones. These were older people who had fled the Croatian city of Vukovar in 1991 and were still looking for a home. Under the Dayton peace plan, they would attempt to go back to Vukovar.

The olive trees, however, were transformed. Someone skillful had pruned them not long before. Competing shoots had been torn away from the trunks. Only a few olives were left, high up in the branches, suggesting a recent harvest. Yes, the manager explained, a crew had taken charge of the trees. They sold the olives to the Makarska mill down past the old port. Following his instructions, I found it hidden behind a Pepsi-Cola plant.

It was Sunday and the front office was closed. Out back, small trucks brought a steady supply for the Pieralisi continuous system, which could handle thirty-five tons a day. The latest arrival was Mato Milkovic, a young waiter who had driven seventy miles from Dubrovnik because the local mill was flattened by Serb naval guns.

During the siege of Dubrovnik, people made oil by stuffing ripe olives in burlap bags and smashing them with hammers. "We got maybe ten liters for two hundred kilos," Milkovic said. "It wasn't too effective." Some farmers went back to the old method that a few rural families still use today: they put the olives in stone tubs, add very hot water, and walk on them with specially made wooden shoes.

Now the road up the coast is open. If there is no mill at Dubrovnik, at least people can drive north. Over the season Milkovic planned to shuttle a hundred tons of oblicas and zuticas up to Makarska. "No problem," he said. "We store the olives in seawater for two weeks to a month. They stay fine, no salty taste. It would be better if we could press them sooner, nearer to home. But . . ." He concluded with a gesture I'd come to know well in the former Yugoslavia. It is a slight lifting of the shoulders and the palms: *C'est la guerre*.

Olives were a big part of everyone's life along the Yugoslav coast until World War II rearranged the pieces. Afterward, the treasured old trees played no big role in Marshal Tito's unified state. The central economy paid farmers next to nothing. High-quality oil was sold for hard currency or bartered for machinery. Especially on the islands, families made oil for themselves and their neighbors. Elsewhere, Yugoslavs ate seed oil.

In the late 1980s olives began to make a comeback. New trees were planted. Agriculture agents helped farmers fight the fly and improve their crops. Production climbed in Istria, near Italy, where Greeks brought the first olives, Romans built large presses, and Venetians planted large groves. When Yugoslavia fell apart, independent Croatia needed soldiers, not farmers. Four and a half years later, however, when the guns fell silent, olivemen returned to their groves.

"Olives are coming back again," Karmelo Barbaric told me, inside the mill. At thirty-three, he ran a restaurant on Hvar. He had just planted another hundred trees to expand the grove he inherited from his father.

Five old-style presses still operate on Hvar, but Barbaric preferred to bring his crop to the centrifuge at Makarska. He was late for the ferry, and a friend was all but tugging at his red scarf. But his dark eyes gleamed at the chance to rhapsodize on his oil.

"Oblica," Barbaric said, holding up a single plump olive. "Top gun." He laughed loudly. "Here we have everything we need. Sun, fish, grapes, olives. On Hvar we say a fish swims three times: in the sea, in wine, in oil."

My last taste of Croatian olives was among the blackened ruins of Plavno, a village in the mountain above Knin. Serbs were the worst culprits in the war, from their systematic leveling of Vukovar by artillery fire in 1991 to the cold-blooded massacre of thousands of Muslims at Srebrenica in July 1995. But the Croats also did their share. When the Zagreb army conquered the Serbs' self-styled enclave of Krajina, on Croatian territory bordering Bosnia, soldiers burned, looted, and killed with savage abandon.

The Croats had attacked early in August 1995, and this was four months later. Nearly all of Plavno's 1,500 inhabitants had fled in panic to Bosnia, along with almost every other Serb in the area. They left only the oldest, too weak or too stubborn to go. Looters torched most of the homes. Only the Orthodox church remains intact. Across Krajina, Croats left the churches alone as tangible proof of their ethnic tolerance.

With a photographer friend I stopped to talk with Jeka Rusic, an old woman who was puttering in her yard. Her son Jenko, forty-one, ragged in torn clothes and broken boots, came out to see what we wanted. He lived on Hvar and came to visit from time to time. Typical of Serb villagers, or Croats or Muslims, they brought out something to eat. Jenko poured sweet homemade wine into cracked tumblers. He put out a small pile of tangerines and waited for us to peel them. Someone had gone a long way to a market and spent badly needed coins for them. To refuse them would have been an insult.

One after the other, Plavno's last survivors stopped by and told their stories. Smiljana Boganic, who was seventy-five, watched soldiers shoot her husband while he stood in their doorway. Up the road, at Grubor, Croats had murdered seven people. Armed looters still came scrounging, Jenko said. As if on cue, two tough-looking young men drove by in a car hauling a trailer. They eyed us carefully and moved on.

Everyone was desperate to leave for Serbia, but no one had travel documents. The Croatian authorities promised them—any day now. Meantime, they had eaten through what winter reserves the looters had left behind. The little money they had hidden away was worthless; Croatians did not want Krajina currency.

As we talked, Jenko went inside and came back with the family's last real food: a sackful of olives he had grown in Hvar. We ate only a few and stopped. Jenko insisted, but so did we. His mother would get over our rudeness sooner than her hunger pangs. Those olives of war would keep Jeka Rusic alive.

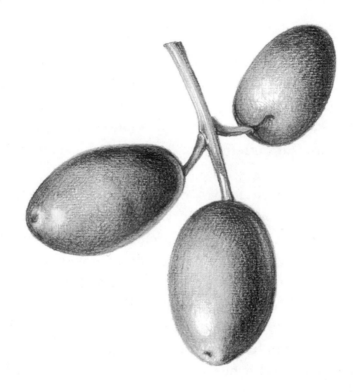

Olive Cluster

Lemon Blueberry Muffins

Arlene Wanderman loves to bake with olive oil. This recipe is from a collection she assembled for *Olive Oil: A Guide for Culinary Professionals*.

½ cup mild olive oil
¾ cup granulated sugar
   Rind of 1 lemon, grated fine
1 large egg, lightly beaten
1 cup low-fat buttermilk

1 to 1½ cups unbleached, all-purpose flour
⅓ cup fine cornmeal
1 teaspoon baking soda
¼ teaspoon sea salt
1 cup blueberries

Preheat the oven to 375 degrees. Grease 12 muffin cups with olive-oil spray or line with paper cups. In a medium-sized mixing bowl, combine the oil, sugar, lemon rind, and egg. Stir in the buttermilk. Whisk the dry ingredients together and add to the wet ingredients. Stir in the blueberries. Spoon the batter into the prepared cups. Bake in the middle of the oven for 20 minutes, or until a cake tester comes out clean. Let cool in the pan on a rack for 5 minutes. As an optional glaze, mix 3 tablespoons of lemon juice and a heaping tablespoon of granulated sugar in a shallow dish and dip the top of each muffin into the glaze while still warm.

Makes 12 muffins.

"Olive, cure all my ills."
—Folk saying, in various forms,
in Spain, Italy, France, Greece.
And, most likely, elsewhere

"Speaking one day of an Aixois who had squandered
his wife's dowry, only Cézanne was not indignant. But,
the victim's parents demanded, can you name one saving
quality? 'Yes,' replied Cézanne, 'I find that he knows
how to buy olives for the table.' "
—Ambroise Vollard

14

The Noble Ingredient

For years now the olive has been worshipped as lord of a holy trinity, along with wheat and grapes, in a religion called the Mediterranean Diet. Its message is simple: By eating like a Cretan farmer, one lives a long, healthy life. That Armenian farmers do all right on yogurt is beside the point. Proselytizers hurl terms like "antioxidant" and "free radicals" like thunderbolts from the mountain.

As with most matters of faith, the fundamentalists sometimes overdo it. Heretics have their own learned killjoys who caution against too much enthusiasm. Those Cretans, for instance, seem to beat heart disease largely because of alpha-linolenic acid, of which little is found in olive oil; soya-eating Japanese fishermen on Kohama Island do just as well.

Unquestionably, olive oil is good for one's health. And to many of us converts, it does not matter how good. There is a more attractive feature than outliving Cretans. Most religions require that a believer die before finding out whose vision of paradise is accurate; the Mediterranean Diet offers its pleasures on earth.

In recent years, medical science has put its stamp on what doctors have known since Hippocrates: Olive oil is easily digested. It helps fight against breast and other cancers. It helps children grow and slows aging. It is good for the bones and joints, the skin, the liver, and the intestines. It fights diabetes and peptic ulcers. But its principal value as a staple in the kitchen is what it does for the heart and the bloodstream: Olive oil protects high-density lipid cholesterol (HDL), the good kind, and it reduces the low-density LDL, its killer twin.

Champions of the olive are now appearing from all corners. For years, back in Tucson, Arizona, friends of mine had sworn by a great bearded bear of a man named Andrew Weil. A graduate of Harvard Medical School, he first made his reputation studying psychodelic pharmacopoeia in Mexico and beyond. Then he taught integrative medicine at the University of Arizona. Private patients lined up to see him.

Weil's book, *Spontaneous Healing*, shot up the bestseller list in 1995, and it settled in to stay. The first line of his eight-week program for "optimal healing power" is: "Go to your pantry and refrigerator and remove all oils other than olive oil."

Weil wrote: "Olive oil appears to be the safest of all edible fats. The body seems to have an easier time handling its predominant fatty acid, oleic acid, than any other fatty acid. Replacing saturated fat in the diet with olive leads to a reduction of bad cholesterol (whereas replacement with polyunsaturated vegetable oils lowers good cholesterol as well)." Weil's conclusion is clear enough: "If the only change you make in your diet is to replace butter and margarine with olive oil, you will have made a tremendous step toward better health and healing."

Weil's advice is hardly new. In 1891 Dr. P. I. Remondino told the Olive Growers' Convention in Sacramento, California: "The modern American, with all his patent contrivances . . . will never know . . . a full tide of health until he returns to the proper admixture of olive oil in his diet. Until he again recognizes the value and use of olive oil, he will continue to drag his consumptive-thinned, liver-shriveled, mummified-skinned and constipated and pessimistic anatomy about . . . in a vain search for health."

The Seven Countries Study, begun in 1957 by Dr. Ancel Keys of the University of Minnesota, is a landmark in medical science. By 1980 Keys's team had compared 12,763 men aged between forty and fifty-nine, in sixteen different test groups. Fatal heart attacks were twice as common among the Italians than the Greeks and Dalmatians. But the rate in Holland was almost double that in Italy. And the Finns suffered 147 percent more coronaries than the Dutch.

"The explanation," Keys concluded, "is the dominance of olive oil in the diet." Oil may not cure heart disease, but it prevents it.

In 1986 Dr. Scott Grundy published more specific findings in the *New England Journal of Medicine*. Grundy is a cholesterol expert at the University of Texas Southwestern Medical Center in Dallas. His impeccable credentials include a string of national association titles and respected publications. A tall, serious-minded Texan, he does not joke about lipids. His study stood unchallenged for nine years.

Grundy's premise was that Greeks and southern Italians eat high percentages of fats but have low levels of plasma cholesterol and not much coronary heart disease. He looked at the effects of olive oil and other fats on cholesterol. Researchers studied eleven men and one woman between the ages of forty-nine and sixty-nine, divided into three groups. After three months, those on a 40 percent fat diet, mostly olive oil, lowered their total cholesterol by 13 percent. Some lowered their LDL by 21 percent without reducing HDL at all.

In 1994 Dr. Christopher Gardner of Stanford University came to an American Heart Association meeting in Grundy's hometown with a different message: People had little to gain from switching to expensive olive oil. Polyunsaturated oils from corn, soybeans, and sunflowers, among others, could be just as effective in keeping LDL down and HDL up. Citing fourteen separate studies, he said that the real danger was the sort of saturates found in animal fat, coconuts, and hardened vegetable oil.

"If you replace the saturated fats you get from meat, eggs, and dairy foods with unsaturated fats from plants and vegetables, you'll be in good shape," Gardner said. "It doesn't matter whether they are monounsaturated or polyunsaturated, as long as they're not saturated."

Grundy conceded the point, but only if polyunsaturates made up no more than 7 percent of daily calories. Beyond that, he told *The New York*

Times, they could have an adverse effect on HDL. Polyunsaturates, he said, are more likely to promote oxidation of LDL, which allows it to be deposited on blood-vessel walls and clog the arteries. Also, Grundy added, high intakes of polyunsaturates in animals impair the immune system and increase the incidence of cancer. "Given this choice," he concluded, "monounsaturates are preferable."

What with mealtimes and snacks in the United States, every third calorie eaten is a fat. Grundy reasoned that unless Americans sharply cut their fat intake, they must necessarily eat a lot of polyunsaturates and monounsaturates. All things considered, he maintained, monounsaturates are better. Especially olive oil.

But Dr. Walter Willett, the olive-oil lover's guru, went further than that. Fats are not so bad, he insisted. And if people stick to olive oil, they are perfectly safe eating 40 percent of their calories in fat. Hardly a quack, Willett is chief of nutrition at the Harvard School of Public Health, with an enthusiastic following of colleagues around the world. He is a runner, thin as a chopstick and boyish behind his bushy gray mustache, a perfect model for his message.

"Eating a lot of the right fats may be better for you than not eating enough," he said, cheerfully adding that this was his biased view. But the mainstream seems to be falling in line. "There are cracks in the 'low-fat' position, and it is beginning to crumble."

Willett said that cutting fat too drastically lowers all cholesterol, which is not healthy. He said the low-fat craze in America stems partly from commercial hype by food marketers and partly from a belief among health authorities that the public is not bright enough to distinguish a good fat from a bad one.

He stressed these points in 1993 at the International Conference on the Diets of the Mediterranean in Cambridge, a grand congress of scientists, historians, chefs, and appreciative eaters sponsored by the Harvard School of Public Health and a food technology think tank called Oldways Preservation & Exchange Trust. He said Americans eat fats because they like them; dire warnings by bran munchers would not make them change. They were better off relying on a healthful fat: olive oil.

Dr. Dean Ornish declared himself appalled at Willett's figure of 40 percent fat in the diet. Ornish, director of the Preventive Medicine Research Institute at the University of California in San Francisco, has reversed heart disease with a no-fat regime. Olive oil is no magic bullet, he said. People are better off cutting all fats to 10 percent of their diet. Like Gardner in Dallas, he attacked quantity, not concept. Olive oil is preferable, as fats go. The real culprit is fat from meat.

Yes, Willett countered, but Ornish also exercised his subjects, and they gave each other group support. He stuck to his numbers. Besides olive oil's effect on heart disease, Willett added, there is evidence that it also fights against cancer. It is rich in antioxidants, which neutralize the free radicals in the bloodstream that can damage cells and lead to tumors. A study among nurses also found that those who ate more antioxidants had a lower risk of heart attack.

In any case, Willett concluded, there are "centuries of epidemiological evidence about the benefits of olive oil."

Europeans are less prone to quibble. An extensive study called "Olive Oil and Health," by Italian medical researchers Publio Viola and Mirella Audisio, reported that olive oil reduced not only heart disease but also gastrointestinal disorders. It helped the natural rhythms of digestion, reduced excess bile acids and free radicals, and stimulated the pancreas.

"Owing to its balanced composition, olive oil has a protective effect upon the arteries, the stomach, and the liver; it promotes growth during childhood and extends life expectancy," the researchers concluded. Being Italian, they added: "At the same time its organoleptic characteristics produce a pleasing sensation to the palate."

In a book entitled *L'Huile d'Olive*, Dr. Bernard Jacotot, a French circulatory specialist of world renown, catalogued olive oil's health benefits. Adding his own findings to others', he concluded that oil in the diet wards off arteriosclerosis plaque and clotting in the bloodstream. It lowers blood pressure. Also, he added, it helps digestion and liver function, builds bones and retards aging, stabilizes glycemia and works against diabetes.

Late in 1995 a Frenchman in gold-rimmed glasses and a white lab coat weighed in with an impressive body of work. Professor Serge Renaud, one of France's leading epidemiologists, published *Le Régime Santé*, a book based on his earlier article in *Lancet* entitled "Mediterranean alpha-linolenic acid-rich diet in secondary prevention of coronary heart disease."

Renaud is one of those silver-haired scientists of the sort only France can produce. With the sober mien of a *président–directeur général*, he emerged from a system where doctors advance in standing like priests toward the papacy. He works in Lyon with INSERM (Institut National de la Santé et de la Recherche Médicale), a government-funded institute not known for slipshod research.

Renaud's team assembled 605 volunteers under seventy who had suffered one heart attack and risked undergoing another. Half followed a conventional medical diet: virtually vegetarian and high in polyunsaturated fats such as sunflower oil and margarine. The others ate like Cretans: fish and poultry instead of red meat, bread, green and root vegetables, daily fruit, and olive oil.

The five-year study was stopped after the fourth year. People on the conventional diet were dying at a rate almost six times higher than the others.

Surprisingly, doctors found that volunteers from both groups had similar levels of good and bad cholesterol. Other risk factors, such as high blood pressure and obesity, were almost the same. But those on the Cretan diet had 68 percent more alpha-linolenic acid in their blood. This vital fatty acid is found in walnuts and *psares*—purslane—a thick-leaved weedy plant that Cretans eat by the handful. Not much, however, is found in olive oil.

Being French, many of the volunteers refused to eat olive oil as their only fat. Researchers allowed them a rapeseed oil–based margarine. Rape, related to mustard, is the source of canola oil. It is rich in alpha-linolenic acid. So is the soybean oil that is a staple of those Japanese Kohama islanders, Renaud noted, which probably explains why they may have the world's lowest incidence of coronary disease.

Renaud concluded that the effect of alpha-linolenic acid on the blood platelets that cause clotting helped to reduce heart failure. But then again, so did the increased amounts of oleic acid, which is found in olive oil, and other elements of the Mediterranean Diet.

Like all good medical research, the Lyon Diet Heart Study answered old mysteries with new questions. Is canola oil better? Renaud does not say so. Andrew Weil warns that supermarket canola is extracted in ways that deform fatty acids, and rape is heavily laced with pesticides that probably find their way into the oil. Expeller-pressed canola from health stores is expensive. Soybean oil, a polyunsaturate, has other drawbacks.

Pending the next round of evidence, the solution seems simple enough. Eat walnuts if purslane is hard to find. And use olive oil.

At that Harvard conference, Antonia Trichopoulou made the telling points. Greek peasants ate olive oil not because they studied *Lancet* and nutrition charts; it was what they had and could afford. After centuries, it was also what they liked. Today, she said, olive oil remains their crucial component. Greeks eat large amounts of raw and cooked vegetables, rather than meat, because olive oil makes them taste so good.

These days, seed oil has nibbled into the market all over the Mediterranean. But reliable and flavorful olive oil still comes at a reasonable cost. People within range of a supermarket, in Athens or Atlanta, lean toward what tastes the best, especially when its food value is unassailable. As far as olives and olive oil are concerned, that ought to close the case.

Americans consumed 64 million pounds of olive oil in 1982. In 1994 the figure was 250 million pounds. Many people who switched for their health are now happily addicted, and they are converting others. With each new medical study, it seems, the number goes up. But the picture is more complicated than it appears. The total for 1995 slipped a little, and sales dropped when prices rose during 1996. The trend was toward better oil. In consumer-sized packaging, extra virgin gained significantly over refined oil. And pomace oil all but disappeared.

"The long-term prospects for the olive-oil market," pronounced one survey written in 1995 for a food distributor by a cliché-loving analyst I was asked not to identify, "are clouded with uncertainty." Cost had pushed many marginal consumers back to seed oil. Prices were likely to go up and stay up for years. Even if the rains hold steady, and Mediterranean trees produce normally again, other costs are rising. Because of world trade agreements, fewer producers will qualify for subsidies; at the same time, European subsidy levels are likely to drop.

Then again, as of 1996 only one American household in five had tried olive oil. In the South and Midwest, the numbers of kitchens using it doubled in a decade: from 4 percent to 8 percent. The field of potential believers is wide open. Food writers have long since embraced the olive, and doctors are following fast. "If olive oil is to continue its growth," the survey concluded, "it must begin to leverage its key benefits and expand the awareness of those attributes that make it unique."

This is the sort of challenge that makes Arlene Wanderman burn brightly. If the Olive Kingdom has an ambassador in the United States, it is Arlene. She operates from a little midtown Manhattan suite on Third Avenue that is cluttered with half-empty crates of oil, precarious towers of brochures and reports, and lined with floor-to-ceiling shelves of food books. She represents the International Olive Oil Council, although the United States is not a member. Her job is to persuade people to consume olives and olive oil—and to show them how. And this comes naturally to her.

"After my first taste of grilled bread and fresh olive oil, I wanted to throw myself into a barrel of the stuff," she told me. That was when she was food editor of the *Ladies' Home Journal* and a self-described expert on tuna casseroles. "I used to be a dietitian at a New York hospital, with degrees, and now look at me. I'm a serious maniac. I'm a believer."

Bejeweled and dramatic, Arlene has a raspy voice that rises and falls with emotion. Her Rolodex is thick with the numbers of nutritionists, oil makers, and food writers in every corner of the world. Her office is geared toward professionals, but the small staff provides a flood of documentation to anyone who asks. For oleic emergencies, there is a hotline: 1–800–OLIVE OIL. Consultants answer the usual run of cooking and consumer questions, but they also have a listing of nutritionists who can handle the finer points.

Arlene insists that the olive-oil boom in America is no passing fad but rather a revolution that is here to stay. Along with its other uses in cooking and salads, she says, olive oil is perfectly suited to baking. A teaspoon of butter is replaced by three-quarters of a teaspoon of olive oil; a cup of butter equals three-quarters of a cup of oil.

She makes a convincing case. Olive oil's small fat crystal yields even, finely textured, and lighter baked goods. The tocopherols act as emulsifiers to produce a smooth, homogeneous batter which makes cakes with a moist and tender crumb. As antioxidants, they retard staling. And, she maintains, olive oil allows the flavor of other ingredients to come through with more clarity.

In 1996 Arlene's office published *Olive Oil: A Guide for Culinary Professionals*, which offers recipes for everything from spice cakes to brownies to lemon blueberry muffins. Butter hardly makes an appearance.

Down around Wild Olives, families cook with nothing else: steaks, omelettes, the lot. In fact, our own oil intake is so high that it is a health hazard. One of these days someone will be crushed to death by all the bottles tumbling off the shelf.

Against all common wisdom, we keep an oil cellar. After two years, only a little of it escapes the lamps. But our cellar—actually the floor level of a cool, dark pantry with some running room so the cat can chase mice—allows us variety. Like wine, every taste and type has its place on the table.

Every kitchen needs a standard all-purpose oil. Loyalty to Emiliano aside, we like ours the best. Most Provence olives, like the Ligurian taggiasca and the Tunisian chemlali, are picked when mature but before the new year. The oil is gold but not greasy, fruity and full, and gentle, but with a light bite to tingle the taste buds. It is wonderful for salads, pasta sauces, fish, *fruits de mer*, and sautéed anything.

When this oil comes straight from the press, yellow-green and murky with unfiltered olive bits, still sharp from the polyphenols that will soften within months, it reminds you yet again why Mediterranean cuisines are held in holy reverence.

Like the locals, we store our own oil in great bulbous demijohns, covered in wicker. This is not particularly wise, since sediment at the bottom can turn rancid and infuse a nasty taste. But it is tradition. And the oil disappears too quickly to run any risk. We keep refilling a slim, tinted glass bottle with a pour spout which lives by the pepper mill and salt shaker.

At least a dozen other oils share the cellar space. Núñez de Prado and a good Kalamata are among our favorites. Both are perfect where a peppery kick is welcome, as long as it does not linger too long. They are terrific straight, on bread for breakfast, and in sauces where the real flavor of olive has to muscle its way past the spices without overpowering them.

For boiled vegetables, we use Tuscans: a sharp green Chianti for delicate cooked vegetables and a Lucca for heartier stuff like potatoes and beans. My other oils are regional specialties I've found here and there, like that sample bottle of Tunisia oil. These I reserve for dribbling on tomatoes, cucumbers, and sweet red onions at lunchtime.

Regional oils serve specific purposes. A Moroccan *tajine* or a Lebanese *Daud Pasha* or many Mediterranean dishes taste best with the flavor of local oils. The blend of spices can depend on it.

Finally, we stock plenty of run-of-the-mill refined olive oil. Deep-frying tacos or chicken is no job for a decent extra virgin. Apart from the wasted money and the spiritual travesty, the flavor is too strong. I like Carli, an Italian oil blended from dependable components and packaged beautifully. Sasso is easy to find in the United States. Pompeiian, from Spain, is a reliable bargain. Even rectified oil can be expensive, but the corners are not worth cutting. I've bought some supermarket brands and ended up tossing them out.

My collection is a bit excessive, but I've never regretted its cost. A few belts in a bar add up to the price of a memorable bottle of oil. If one is picky about wine, a choice vintage can equal reasonable oil needs for months. It is all a matter of one's priorities. As for nutritional value, Walter Willett and Dean Ornish can duke it out all they want. I know where I stand.

Fearing a taint of prejudice, I consulted my neighborhood cuisine queen, Patricia Wells. We met at a dinner at l'Orangerie in 1979, and I've associated her with wonderful food ever since. So does everyone else; her brisk-selling books are bibles for eaters in France. *Trattoria* is Italian cooking made easy.

"My favorite oil is the Maussane," she pronounced, referring to the Jean-Marie Cornille mill run by Suzy Ceysson. Next she likes other oils from the Les Baux region, near where she happens to live part of the year. "I've tried not to be partial, but I think French olive oil is best. I use it for everything. Even Italian cooking. The only Italian oil I like is from a little place near Florence." She has yet to try a good Kalamata.

Patricia did not think much of my oil cellar idea. "To me, this business of keeping forty different oils is ridiculous," she said. "I just open up one bottle, and when that's gone, I open another. I don't fry much in olive oil. That's too much of an extravagance."

Nancy Harmon Jenkins, my other expert witness, takes a different view. By coincidence, I met her the same year as Patricia, at a restaurant in Rome. Like Patricia, she went from creating high art in her own kitchen to writing cookbooks of the finest sort. Her *Mediterranean Diet Cookbook* is a definitive masterwork.

Nancy recommends at least four types of oils: a mellow all-around oil, Ligurian or French; something fruity and specific, perhaps a Baena or a Kalamata; an inexpensive oil for frying; and a sharp, bitter Tuscan, if you like it. In her own kitchen, she alternates her favorites as the mood strikes.

In fact, every expert argues a different case. Excellent cookbooks contradict one another. Cooking with olive oil, and olives, is an art that is no more precise than pruning. Certain basic principles apply. The rest is personal, based on taste and whim. In general, olive-oil cooking is like tending an olive tree: the more effort you put in, the greater the reward.

The first step is to get properly oiled. For this, forget everyone's advice. Taste as many oils as possible at specialty stores and friends' homes, and then take the plunge. It is hard to go far wrong; you can always massage toes with the least favorite. Try a good Tuscan and then a mellower Italian and some Spanish and Greek. If you can find a good French oil, grab it.

Organize an oil tasting with friends. If everyone brings a different bottle, you can all sample a wide range and compare notes on each. You need only teaspoons, but chunks of unfancy bread, celery sticks, and soda water help clear the palate.

Oil scouting can be a pleasant sport. Almost every urban area has a specialty store that will offer advice, if not a sample. On the East Coast, Dean & Deluca is one among many fine shops. Californians have a wide choice, including Katz & Company in Napa and Corti Brothers in Sacramento. And there are plenty in between. West Point Market in Akron,

Ohio, for example, stocks Frescobaldi Laudemio, Lapo Mazzei's Castello di Fonterutoli, and Lérida from Spain. Zingerman's in Ann Arbor, Michigan, mails oil around the United States.

Extra virgins from the same place will vary from year to year, but the general characteristics and quality are likely to stay the same. Some of the best oils, such as Jean-Marie Cornille or Saint-Côme from Cadiére d'Azur, are sold only at the mill or in a very few places. Or by mail. It is a matter of asking around, or simply noticing small signs at the side of the road.

Olive fairs are the best place to establish a regular supply line. Oleum, held each December, has the advantage of being in Florence. Expoliva in Jaén is bigger, but with a natural accent on Spanish oils. Cruise the stands and sample. When you find a producer who moves you, arrange for regular shipments. He will in all likelihood be thrilled to accommodate you. No maker of good oil can refuse honest adulation from afar.

When traveling in olive country, check local markets and ask around. People keep the best oil for themselves but will often sell you some. If I hear about a secret source, I'll make a detour. I've brought home some of the best oil in whiskey jars or plastic Pepsi bottles. You never know what you'll find, but it's always worth trying. At the least, you'll visit some picturesque old cellars up cobblestone lanes.

You may also want that neutral-flavored pure olive oil for hard labor in the frying pan. Price and fancy labeling are not reliable clues. Trust your own senses. Consult a specialty grocer and take a chance. Once you find a refined oil you like, you can be fairly sure the taste and consistency will not vary the way extra virgins do.

Supermarket chains usually offer their own inexpensive house brands. If nothing else, they are cheap enough to check out at least once. And with luck, you might find one to your taste.

The big brands sell something called "lite" olive oil. The price is higher because people will pay it. In fact, it simply contains less extra virgin, so it is cheaper to produce. It has the same amount of calories, about 125 per tablespoon, as extra-virgin or seed oil. It is useful for people who want to fry in olive oil without any intrusive flavor.

Keep oil in a dark, cool place; light deteriorates it quickly. When a bottle gets low, transfer the oil to a smaller container, well stoppered, to slow oxidation from air. Oil solidifies if too cold, but then liquifies again without damage. Keeping it refrigerated is neither necessary nor a good idea. Water droplets collect from condensation. I have a friend in Colorado, a gifted chemist and ardent gourmet, who fast-freezes the oil he brings back from Italy. This presumably stops the aging process; he knows what he's doing. Still, it takes him a long time for meltdown when he needs a quick splash. Spanish and Tunisian oils tend to last the longest, if properly stored.

For cooking, the basics are simple: Because of olive oil's high smoke point, you can fry with it three or four times without courting carcinogens. Pat Wells does not do this; master chef Joël Robuchon does. But don't switch from fish to, say, eggplant. Flavors and strong spices linger. You can strain the oil with a coffee filter between uses and freshen it with a squeeze of lemon.

Always heat olive oil before putting in any ingredients. A heavy-bottomed skillet will diffuse the heat evenly. Frying is best at around 390 degrees, just before the oil releases curls of smoke; that is hot enough to crisp a small piece of bread in about a minute. In frying, the more oil the better. Hot olive oil seals most foods with a crisp caramel-like film so they do not get soggy from absorbed grease, as with seed oils. If food floats freely in the pan, it is more quickly and evenly sealed. Fry in small batches so the oil's temperature does not drop.

Besides frying and baking, olive oil is perfect for marinating and basting meats to be roasted, broiled, or barbecued. Used hot, it is just right for stir-fry. Over a low fire, it makes scrambled eggs with a subtle new flavor. It is handy for simmering: a tablespoon of olive oil added to rice when it's cooking helps keep the grains separate. In pasta water, it guards against sticking.

Bear in mind that the health effects of olives depend mainly on what you eat with them. Italians, who tend not to consult the University of Texas on culinary matters, love dishes like *olive fritte*, green olives fried with cholesterol: streaky bacon, pork, an egg plus an extra yolk, and Parmesan cheese.

Beyond the basics, you can get as complex as you like. Paula Wolfert offers a range of olive recipes in her classic *Mediterranean Cooking*. Maggie Blyth

Klein, who feeds happy diners at her Oliveto Restaurant in Oakland, has some tasty recipes in a good-natured little book called *The Feast of the Olive*. A much fatter, fuller tome is Sarah Schlesinger and Barbara Earnest's *Olive Oil Cookbook*. I like Mireille Johnstone's *The Cuisine of the Sun*. The list is long.

For fresh ideas, try some of the lesser-known cuisines. A useful book called *Cooking with Olives* is published by the International Olive Oil Council. It offers recipes from thirteen countries around the Mediterranean rim, beginning with Algeria, where French fanciness blends with spicy Arab exotica.

At Wild Olives, Jeannette's specialty is olive pie, from a Greek recipe she dug up somewhere. She chops 1 cup of pitted black olives along with ¼ cup of mint and 1 small, finely diced onion. In a separate bowl she mixes 1½ cups of flour, 3 teaspoons of baking powder, and nearly a cup of olive oil, adding in enough warm water to make a fluid dough. Then she adds the olive mixture and keeps blending. She bakes it in a floured pan at medium heat for about an hour.

Some of the best dishes are the simplest, such as calamari rings, Spanish-style. Salt them, shake them in flour, and fry in lots of hot olive oil. For that matter, Spanish *gambas al ajillo* are even better. Peel 1 pound of shrimps and toss away the heads. Heat 6 tablespoons of oil in a heavy pan and, on low heat, sauté 2 cloves of sliced garlic until golden brown. Flake in half a small dried hot red pepper and then zap the shrimps for a minute. Sprinkle on salt.

Even simpler are potatoes mashed with olive oil instead of butter or milk. If the oil is good, warm it first and add enough so that a thin layer rises above the creamy potatoes. The Greeks pile on garlic.

And there is the stalwart Provençal hero sandwich, *pan bagnat*. Slice a large roll or small loaf in half and scoop some bread from the center. Douse the bread with oil. Lay on thick wheels of ripe tomato, red onion, sliced hard-boiled eggs, and salty anchovy fillets. Sprinkle more oil on top, add the other half of the bread, and wrap the whole thing in a moist cloth. If the sandwich is to go with you on a roam through the lavender, rewrap it in aluminum foil. And don't forget the Bandol.

———

There is only one reliable way to buy olives for the table. Stand in front of open tubs and taste them, one by one. A merchant who won't allow you this simple liberty is in the wrong business. In the United States you may need a little scooper and surgeon's gloves. Anywhere in the Mediterranean, just seize a likely-looking olive in your fingers and pop it in your mouth.

But beware. Names and labels tend to mean little. You may fall in love with an olive only to find yourself on a long search for its equal. In Paris, I once bought a stock of green picholines in hot chili peppers and fresh herbs. The Moroccan merchant called them "cigalo." That sounded like the Provençal word for cicada, and I thought I had fallen onto some useful lore. I hadn't. The name meant nothing to anyone else. When I finally found the same olives at the Draguignan market, the lady selling them chuckled at my story. "People who make olives call them anything they want, and it changes from place to place."

Curing your own olives is a great hobby if you've got the patience. However you do it, some basics apply: Green olives should be cracked and soaked in water, regularly changed, for at least eight days. Black ones should be pricked with a pin a half dozen times or scored with a knife and then treated in salt, dry or in brine. For dry curing, alternate layers of olives with layers of crushed rock salt. For brine, add salt to water until an egg floats in it and cure the olives at least a few months. Not enough salt in long-cured olives allows fearsome bacteria to form. Too much causes the olive to shrivel up in wrinkles.

You can add seasoning at any stage, depending on how delicate the flavor and how deeply you want it to penetrate. Lemon goes well with green olives; chilies and garlic are good with either. Mediterranean herbs are best with black olives, and they need not steep very long. The possibilities are endless.

Nancy Harmon Jenkins buys cured olives and marinates them. One favorite way starts with 1 pound of drained little black Niçoises. She pares and slices into a fine julienne the zest of 2 lemons, discarding the white rind. The yellow flesh is also sliced thinly, with the seeds removed. She tosses the lemon and olives in a bowl along with a dash of cayenne pepper, ½ teaspoon of sweet paprika, and ¼ teaspoon of ground cumin. Then she adds a tablespoon, or more, of extra-virgin oil and tosses it all again. After a day or so, they are ready.

To keep prepared olives for more than a few days, add more oil to cover them. For her black Niçoises, Nancy puts the olives in a jar after sprinkling in a tablespoon of salt. They may be kept refrigerated for two or three weeks. But she serves them at room temperature.

Patricia Wells makes terrific black olives from scratch, dry and in brine. Her place in Provence lies within the designated AOC region of Nyons, where she grows tanches. She picks late in December, when the olives are black and ripe. For both methods, she jabs each olive a few times with a seafood fork.

For black olives in brine, the metric formula is simple: 1 kilo of olives; 1 liter of water; 100 grams of salt. (2 pounds of olives; 1 quart of water; 3½ ounces of salt.)

One batch gets mixed only with *gros sel*, coarse salt, one hundred grams of salt for a kilo of olives. "I keep them in a great big bowl in the kitchen and toss them every time I pass," she said. "In six or seven days they're wonderful. They taste like fresh olive oil." If the race is on for Christmas, she freezes the olives for a day, which cuts the time by a third. That is an old Provençal trick. Wait for cold weather to shrivel the olives a little, making them easier to cure. Patricia often makes the liquid from a nasty-looking mother brine, fifty years old, but uses fresh brine if she does not have enough. The olives sit for two months.

Patricia seasons the olives at the end with homemade *harissa*—Tunisian chili paste—garlic, or rosemary. "By the time we get tired of eating the dry-cured olives, of course, the others are done," she said. "Isn't nature wonderful?"

Olive Blossoms

Wilted Cabbage Salad with Melting Blue Cheese

Vegetables—fresh, cooked, or halfway in between—come to life with olive oil. Peggy Knickerbocker, a San Francisco writer and olive lover, likes this salad.

⅓ cup extra-virgin olive oil
6–8 slices pancetta, or bacon, cut into bite-sized pieces
1 medium red onion, finely chopped
4 cloves garlic, minced
1 small red cabbage, cored, outer leaves discarded, sliced very thin

1 teaspoon fennel seeds
6 tablespoons balsamic vinegar
Dash Worcestershire sauce
Salt and pepper to taste
⅓ pound blue cheese or feta, crumbled
⅓ cup flat parsley leaves, chopped

In a large heavy skillet, warm the oil over medium heat for 1 minute. Add the pancetta, onion, and garlic. Cook for about 10 minutes, or until the pancetta has crisped and the onion is transparent to golden. Add the cabbage, fennel seeds, vinegar, Worcestershire sauce, salt and pepper. Toss to combine the ingredients so that all are lightly covered with dressing. Taste for seasoning. Transfer to a shallow serving dish and sprinkle with the cheese and parsley. Serve warm or at room temperature.

Serves 4–6.

*"Olive trees are angels
anciently wakened from the sky
under the earth."*

*"Gaze long inside the mind
at the black fire in the olive."*
—Willis Barnstone

*"The perfect capsule
of the olive
filling
with its constellations the foliage;
later
the vessels,
the miracle,
oil."*
—Pablo Neruda

15

California Olive Rush

The California olive-oil renaissance took off in the early 1990s, not long after Bruce Cohn's wife decided she had had enough of those black splotches he tracked across her carpet every year. Do something with those damned trees, she told him, or cut them down. Cohn was making pretty good wine near Glen Ellen, and those damned trees were why his vineyard was called Olive Hill. The stately old picholines had come from France last century with a long-forgotten Sonoma settler. So Cohn decided to make some pretty good oil.

This was no farfetched idea. Junípero Serra and the Franciscan fathers had brought trees from Mexico to the Mission San Diego de Alcalá in 1767. Along with Catholicism, they implanted olive culture up the coast, beyond San Francisco. Beautiful old trees still ring the mission at Santa Barbara. In 1803 a missionary named Father Lausen reported that California oil was excellent. By 1885, produced commercially in Ventura, it ranked with the world's best.

New varieties from Spain and Italy added to quality and taste. Early this century, however, Italian clans flooded the American market with cheap oil from the old country. Local growers found a way to make their olives into tasteless black lumps in cans.

Huge plantations for the canning industry grew up in the Central Valley. But in Sonoma and Napa, great trees dating back to the state's early history were dug out to plant vines or sent south to decorate parking lots around Los Angeles. And then late in the 1980s, when a growing number of Americans began to realize what olive oil could do for them, Californians again noticed the trees that remained.

Cohn's lineage was neither Mediterranean nor oleaginous, but he was a born hustler of the charming kind. He grew up in Chicago, the son of Sam Cohn, a Jewish shoe salesman who sang Italian arias at the opera under the name Roberto Conati. Having made some money, Bruce entered the California wine aristocracy the way most people do, with a down payment and a mortgage.

Around Halloween in 1992, Bruce and his brother, Marty, trucked their olives down to a mill in Modesto, three hours south. B. R. Cohn extra virgin was not bad, but its posh hand-etched French bottle was terrific. The first crop of two hundred 12-pack cases sold out before the next picking. At $50 a half liter, Bruce's amateur oil from unkempt trees, pressed off the premises, cost more than twice as much as Laudemio.

Lila Jaeger of the Rutherford Hill winery over the hill near Napa had been pressing her olives since 1989. A handful of other Sonoma and Napa vintners were on the verge of commercial production, or were thinking about it. Bruce's success removed all lingering doubts.

I chanced upon Olive Hill one morning on a drive up Highway 12 from my sister Jane's house in Sonoma. Just past M.F.K. Fisher's beloved little cottage, I suddenly applied the brakes. A thick grove of towering olives swept back from the road toward an elegant home that looked inhabited by people who would not appreciate uninvited callers. The hell with it, I thought; reporters are supposed to be nosy. At the end of the driveway I was directed back down the road to a public back entrance which led to the tasting room.

As it happened, this was the first year of production. The extra virgin was very good, nippy up front, smooth, with a clear green-yellow color. Picked early, it had a harder bite than French picholine. I liked it. But $100 a liter? Cohn also offered a second, lower-priced line of oil from olives he bought around the neighborhood, a nondescript mélange that might have been better steeped in herbs. It comes in a similar 14½-inch nearly black bottle, but the label is etched by machine, not by hand.

When I announced my purpose, someone activated the intercom and Bruce shambled out of a back building. He was handsome, squarely built, and supremely self-confident, the sort of guy a loan shark might dispatch to collect a bad debt. His lazy eyelids did not conceal a Dennis the Menace gleam in his eye that suggested he was cooking up something you'd hear about later on. He laughed easily about his new role as oil baron and seemed happy with the early results.

I dropped in a year later and found Cohn in a barn, smoking a foot-long Cohiba cigar and contemplating a new challenge: a four-wheel-drive hot rod. He had shoehorned a souped-up Toyota engine into a rebuilt Willys jeep chassis. The oil venture was thriving, in the hands of Greg Reisinger, his partner, an immensely pleasant management expert with a sandy blond pigtail who had taught himself the olive business overnight.

Reisinger had been elected president of the California Olive Oil Council, which by the end of 1995 had 175 members. About twenty-five were active, and a handful were selling at least some oil. The group had taken shape under Lila Jaeger of Rutherford Hill as a friendly circle where fledging oil makers could compare notes on success, set standards, and commiserate over setbacks. They had traveled to northern Italy and southern France, and they worked together to promote California oil.

Olive Hill had begun to host the annual California Olive Oil Festival in June for all producers. It draws thousands of eager dunkers and slurpers. The Cohns' own tasting room was about to take over the main house; Bruce was planning something better for his family. The estate command post had spread out back into a suite of computer-studded offices, paneled on every wall with dozens of gold and platinum records. Bruce's night job is manager of the Doobie Brothers.

"We think California should produce a world-class oil of its own," Reisinger told me on my third visit, late in 1995. "I don't think our goal should be to replicate Italian oil." From anyone less likable, these might have been fighting words. A main thrust of the California olive renaissance is to import high-strung frantoios and leccinos to produce high Tuscan green oil.

"I believe Americans need to be educated to distinguish the different oils, to learn more about them," he said. "American food preparation ought to be simple, elegant, and healthy, and we believe olive oil falls under this."

The problem, he acknowledged, was that no one could quite work out what kinds of olives should go into a California oil. After touring Europe, he found he preferred French oils. "They are more delicate, without that sting at the first bite." He held up a bottle of Fabrice Godet's Saint-Cassien from the old stone mill in Draguignan; he was my neighbor who worshipped the Núñez de Prado brothers and made oil that was in a class with theirs.

Reisinger also liked Madame Allione's oil, and he'd come away with the same impression I'd had. "She's a great marketer," he said. "You get the idea that if suddenly lavender oil was the big thing, she'd switch to that." Small world.

Cohn agreed, and he was planning on buying new trees from France. As it was, Olive Hill depended on its single French picholine variety. Lila Jaeger had just ordered three hundred bouteillan and aglandeau from a nursery just down the road from Wild Olives. They would join her two hundred century-old Spanish trees. Several other vineyards were doing the same.

But the two biggest players had started from scratch, each importing thousands of Italian seedlings as a first step. Both plan to sell trees to smaller producers seeking to join the wave. A North Beach Internet software entrepreneur named Ridgely Evers, who also invented a bicycle lock, had just produced a first crop of DaVero, an intensely sharp and green oil, from his three thousand young Italian trees.

"The challenge for California is to step all the way back and develop a selection of varietals," Reisinger pointed out. "There is a real market out there, but the biggest problem is to supply a consistently high quality and quantity. If people back East hear stories of bad California oil, that will hurt us for a long time. We need a world-class oil. That's a lofty goal, but the right one. If we stop short at anything else, why bother?"

For California producers, that means a tough uphill climb. Most trees, overirrigated and underpruned for generations, grow olives that are largely water. Cohn's three hundred picholines seldom yield better than 5 percent. And the five main varieties are better suited for table olives than for oil.

The Franciscans' original trees were missions, hardy Mexican variants of a Spanish tree that came to the New World with the *conquistadores*. Mission olives predominated in California until 1875, when Andalusians brought the manzanillo. Ten years later Spanish immigrants planted the sevillano, the fat gordal still popular around Seville. At the same time, Italians imported the ascolano. Then, in 1905, growers added the barouni from Tunisia. Beyond those, at least sixty other varieties are scattered around the state, from the Redding picholine to the Aghiza Shami from the vicinity of the Egyptian pyramids at Giza.

A few experienced California oil makers produce good blends of mission and manzanillo. But they are lucky to get a 10 percent oil content, half the average Mediterranean level. Early in the season the yield is closer to 6 percent. That means a lot more olives must be pressed. And all that vegetable water washes out flavor.

"We're sort of learning as we go along," Cohn said. "But the results are encouraging, and people seem to like our oil. We're going ahead with this."

First he realized that to pick olives you must get your fingers somewhere in the general vicinity. California law forbids workers to climb ladders higher than eight feet, which was about where his trees just got started. Their energy went to old thick limbs, soaring skyward. Little of the olive-bearing new wood grew each spring.

From all indications, Cohn and Reisinger were learning fast. Both peppered me with questions about how different cultures made oil; they were among the few California olive people who showed such curiosity. More often, people repeated with confidence the predilections of a few popular European gurus. The first pruning job at Olive Hill was an unceremonious beheading of trees that had soared out of reach. Untrained new growth burst forth in every direction, leaving the crowns a tangled mess.

Now they are beginning to take shape, and their crews are learning how to pick.

They also realize their limitations. Rather than press their own oil, they still go the Modesto mill, run by old Joe Sciabica, whose father came from Sicily in the 1920s. The Sciabicas have been making their own fine oils for sixty years, and they are happy enough to press other people's as well. "Any one of us could learn to run that machine," Greg said. "I could stand there and ask a million questions and follow the same procedures. But making oil is an art. They're a little modest about it, but it is in the family. You don't get it by going to training classes. You just know. I just feel very strongly about that."

The 1995 B. R. Cohn estate extra virgin was excellent oil, especially to anyone for whom $50 a half liter is pocket change. Apparently, that is no small group. Cohn's supply sold out fast, and he is scrambling to find all the room he can for new seedlings.

"We'll get there," he said, shooting his associate a self-mocking grin. "We're just sort of going slowly, taking it tree by tree in the forest of life."

Lila Jaeger had no carpet problems. Her story reminded me of Wild Olives back in France. She was clearing away a jungle of oaks and manzanita on a back piece of her property, and she struck olives. They were nevadillos and a rare California cultivar called cornezuelo. When they recovered from the bushwhacking, her trees offered a glorious crop of fruit. Lila loved the idea of producing oil, so she did it.

"It was pretty hit-and-miss," she told me, acknowledging the understatement. Her trees suffered from a savage pruning. Changeable weather did not help. She picked once in early fall, and another time in early spring, trying to find the right balance. By 1995 she had found her pace. Lila Jaeger Extra-Virgin had its following.

Lila is the picture of a Northern California gentlewoman olive farmer, outdoorsy and elegant. We sat in a sunroom of her rambling house in St. Helena, an all-seasons greenhouse of succulents and flowers. Her library was stacked with books on the subject.

She has led the fight to keep California oil respectable, insisting that high standards be enforced by the council. At one meeting Greg Reisinger argued that "estate-bottled oil" could contain 50 percent of oil from other places. His idea, he explained to me later, was that producers could buy oil from people who had the same variety of tree. "We told him, 'We don't think so,'" Lila said with a belly laugh. "He shrugged and went along with us." The council's estate-produced limit is now 95 percent.

Lila had a light year in 1994 and another in 1995. "We had unseasonal rain at bloom time," she explained, with the matter-of-fact air of a grower getting used to such things. "We'll have maybe four hundred bottles, eight hundred at most." She uses a simple 375-milliliter flask, half the size of a wine bottle. Her preferred retail price is about $15, compared, say, to $25 for Ridgely Evers's green oil.

She was still experimenting. With too few olives to meet Joe Sciabica's two-ton minimum, she decided to try a traditional-style press. "It's like the old method of making wine," Lila said. "Often you got wonderful wine, but often you didn't. I lost a quarter of my olive crop last year—I took them to a guy who'd bought Tinkertoy equipment."

Eventually Lila hopes to turn her passion into something lucrative, but she seems pretty relaxed about it. "It takes time and effort," she said. "It's just like the wine business. If you don't keep stirring the pot, it's going to go away. There's a small market at the high end."

Down the road in Napa, Albert and Kim Katz's shop stocks the best of California oils. Albert has his own strong opinions, but he will pour out samples for anyone interested enough to ask and patiently explain the nuances. This is a valuable service. In the full flurry of the California olive rush, you can't tell the players without a program.

Only a few producers bottle oil from their own trees. Others buy olives carefully from estates they know, blending neighbors' oil with their own for a relatively consistent taste and quality. And a few, with no trees at all but unwilling to miss out on a good thing, buy mediocre oil from elsewhere in the state, which they sell under their own name at ludicrous prices. Since the United States is not a member of the International Olive Oil Council, no organoleptic panel tests are required. Anyone can declare almost anything "extra-virgin."

Albert Katz has no illusions about Napa and Sonoma oils, but he is optimistic. Some are good and will get better as young trees age and older ones are tamed. Producers are learning which seedlings to plant and when to turn off the irrigation. Among others, he likes Soda Creek Ranch oil from picholines and Kimberley, the only California oil stocked at Dean & Deluca. I was intrigued by Olio d'Oro, a fresh and peppery oil, more green than gold, which the Harrisons press on their mountaintop estate off the Silverado Trail, not far from Napa.

When I phoned Jill Harrison, the family's oil maker, she was just going out. Ask for Pablo, Jill said. "But don't go to the main house," she added. "You might wake my father, and he'll be mad." I approached cautiously, trying not to crunch gravel and disturb any ogres lurking in the house. A man stood in the drive, swarthy, unshaven, and young-looking, a likely mill hand. "Are you Pablo?" I asked. "No," he said, "I'm Michael." As it turned out, Michael Harrison, Jill's father, was neither angry nor an ogre. After an amused "And you are . . . ?" he showed me around.

Harrison, who was fifty-seven, had made a ton of money as an international marketing consultant, and he bought his vineyard overlooking a man-made lake and the Napa Valley beyond. He had some old trees. The family decided to buy a mix of varieties from other estates and make oil in the barn. Jill would be in charge.

"We went to Italy and looked all over for weeks for an old press," Harrison said. "Finally, we found this in somebody's back yard." It was a hundred-year-old pint-sized traditional press with small mats; like the Italians, Californians call them *fiscoli*. It was from Montefiridolfi, near Fonterutoli, by Siena. "The thing came with all the instructions in Italian, and we had a hell of a job putting it together."

He also bought a $20,000 crusher, with two granite wheels that weigh two tons each. The stone surfaces are smoothed every other year by the tombstone maker in Napa. The Harrisons press with nylon mats, steaming them regularly and freezing them at night to halt oxidation. After a few seasons, they invested in a centrifugal separator. It finished the job more cleanly and quickly. They now turn out four hundred cases a year of Olio d'Oro. Michael does not contend it will cause Tuscans to seethe with jealousy, but it is fruity and fresh, and a lot of people like it.

"We have fun," he said. Before settling into Napa, the family almost bought a place in Lucca. "We like the tradition, the thoroughness of the process. If I didn't make this oil the right way, I might as well go back to New York." He goes back a lot, anyway, he said, with the masochistic snort of a true New Yorker. "I need a little tension in my life."

He showed me his front terrace. A valley of vines and lush trees spread out in a pastel panorama beyond the lake at the foot of his mountain. Out back, we talked above the chatter of seven hundred tropical birds. His neighbor Gloria Allen is a specialist in returning rare birds to the wild. The other neighbors only make wine.

Harrison did not take himself too seriously and offered a few amusing swipes at the new olive people who did. But he was happy to see people bringing back old trees and making decent oil. "It's coming along," he said. "We're trying to talk it up as much as we can."

The McEvoy Ranch is tucked back in the rolling hills between Petaluma and the Pacific, well to the west of Harrison's lush mountaintop. It was an open-range dairy farm back in the days when northern Marin County had room for such things. Today, if you squint a little and ignore the scrubby chaparral, it is the lowlands near Lucca. Neatly spaced young olive trees climb up the steep slopes and disappear from view. Nan Tucker McEvoy, *grande dame* of San Francisco society, has another vision entirely than the winemakers who dabble in oil as a sideline. She wants to make quantities of Tuscan-type oil in her own little Tuscany.

By the main house, elaborately cut stone walls demarcate the lush and colorful gardens. In a large greenhouse long tables of potted cuttings wait to be planted so they can produce the frantoios, leccinos, and pendolinos so dear to the Lucchese.

"I bought this place because I lived twenty-five stories up in a condominium in the city after thirty-three years in Washington, D.C., and I wanted somewhere in the country to get away," Nan McEvoy told me over a lunch of vegetables and oil. "My grandchildren did not want me to be too far away, so I found this." I had guessed that price was not a consideration. Despite her jeans and workshirt, she was no farmer. A de Young of Nob Hill nobility, she is the closest thing California has to a Bona Frescobaldi. Among other holdings, she owns a respectable interest in the *San Francisco Chronicle*. She gives vast sums to the arts. Nan McEvoy is, in a word, rich.

Mrs. McEvoy's grandfather Michael de Young, "the General," founded the *Chronicle* in 1865. In the early 1990s, as chairwoman of the board and dominant stockholder, she fought a bitter battle with family members who wanted to sell the paper. But in April 1995 she was deposed in a coup d'état. Her rivals passed a bylaw that directors had to retire at seventy-two. She was seventy-five. She left the boardroom to enjoy life at a more peaceful pace.

The 550-acre ranch was zoned for agricultural purposes, and Mrs. McEvoy could not build a house without producing something. "I could not see myself raising dairy cattle, and I liked olive oil," she said. "So here we are."

As guru, she hired Maurizio Castelli, an erudite Tuscan with a black walrus mustache, beloved in California olive circles. Shari Gonzales, an enthusiastic young agronomist, supervised the planting. By the end of 1995, 5,200 trees were in the ground and more were on the way. Oil production would not be far behind.

It was an enormous amount of work. The clay soil needed major drainage to satisfy the temperamental trees. Rows are not terraced, but are reached by access tracks that were carefully planned to prevent erosion. Irrigation lines were laid from the ranch's lake. Shari put in vegetable gardens and tall flowering trees. If you happen upon it suddenly in the Petaluma hills, the sight can stop you dead in your tracks. Though Mrs. McEvoy still lives in the city, thirty miles south, she describes each visit as a new thrill: "Every time I come around the corner and see an olive tree, I say, Gosh."

The trees were flown in from Italy, each wrapped in plastic to keep moisture around the young roots. Regulations permit trees but no foreign dirt. When the first thousand arrived, the customs inspectors wanted to see each one. But it was the start of the Thanksgiving holiday. "You can imagine how hard an inspector works on Thanksgiving," Mrs. McEvoy said, her face set in mock horror. The trees finally cleared, but their tags were jumbled and no one knew which variety was which.

Ridgely Evers had it worse. He had arranged to truck his seedlings to Frankfurt for a direct Lufthansa flight to San Francisco. The Italian shipper did not think it proper for Tuscan olives to fly aboard a German carrier. Instead, the trees flew from Milan to Rome and then, aboard Alitalia, to Los Angeles. Evers had to hurry down to clear his bedraggled trees in person. U.S. Customs inspectors had found a speck of dirt on one of the young trees, and he had to clean every one in the shipment.

The system is smoother now, but the authorities are tough. No imported tree can be sold for two years, a sort of house-arrest quarantine. If a tree dies, its corpse must remain in place until agriculture agents inspect it.

"We're selling trees, but now only a few," Mrs. McEvoy said. "We're still learning how to make cuttings for the nursery. It's a bit of an art form."

Mrs. McEvoy is sanguine about her venture. "It's very stylish to be in olives now," she said, chuckling. "Like everything else, it will pass." But she intends to stay with it. A lifelong butter person, she was converted to olive oil by Marcella Hazan, the high priestess of Italian cooking. "I wasn't anti or anything. I just didn't realize what it was." Later, Maurizio Castelli hooked her on the agricultural side. "I love it all," she concluded. "When you pick olives, that's mystical stuff."

Across the table, Pia McIsaac laid out a more down-to-earth vision of an olive empire. An Italian of Scottish bloodlines, she was overseer of the McEvoy ranch. She planned to sell seedlings to Californians who want to restore old groves. When all those trees start to grow, the sky is the limit. Relative inexperience, she said, would be an advantage. "We can get all the knowledge from other oil nations. Without all that trial and error of so many generations, we will have the possibility of really competing in the international market."

She held out her arms to make the point. "We are here now," she said, waggling her left land down around her knee. "Soon we will be here." Pia McIsaac's right hand, at shoulder level, was a little higher than others might have placed it. California did it with wine. Why not olive oil?

A few months later, she was no longer employed at the ranch. Mrs. McEvoy was happy that the grove had taken roots and that the first oil had come off the presses. But she was in no hurry to build any empires. She was in it for that mystical stuff.

For all the talk about oil, 80 percent of California's crop is for "pizza olives," those tasteless black slices that lurk in salad bars and free-delivery delis. That includes canned olives, which are graded in size from large to gargantuan with a thyroid problem. Some green olives are produced, but most of those come from Spain or Morocco. The state's specialty is the California ripe black olive, which is neither ripe nor black. It is picked green, for firmness and skin texture. Lye quickly leaches out oleuropein. A bubble bath of ferrous gluconate turns the olive black. Flavorless brine embalms them for an almost indefinite shelf life.

Jim McGovern, who runs a food import company called I-MAG in San Francisco, tries hard to push olives flavored with laurel, thyme, and rosemary, treasured delicacies in Europe. "If it wasn't for the ethnic population, we wouldn't sell many," he said. The market is in isolated pockets, such as Detroit, which has the highest concentration of Arabs in America. It is growing in California among urban food lovers. "You have to have been to Spain or Morocco, or have been turned on to good olives by friends. That does not amount to a lot of people."

Even olive oil is likely to remain a luxury item in the United States, which lags far behind Europe in appreciation of the noble fruit, McGovern said. He is not your run-of-the-mill food merchant. Earnest and widely read, McGovern has the air of Peace Corps idealist gone mainstream. He has seen the world, developed a passion for one of its microcomponents, and found a way to nurture it while it makes him a decent living. His passion is olives.

At school in Caen, France, McGovern met a Moroccan student who has been a close friend ever since. Later, at the Marrakesh market, he saw the light. He traveled in North Africa and Spain to learn what he could and then set up business in San Francisco. His original International Market Advisors Group was renamed I-MAG when it shifted from advice to imports. "Now," McGovern says, "the only consulting we do is to convince our customers to buy from us." He handles a lot of food items, but people know him as an oliveman. Along with table olives, he sells bulk quantities of high-quality Spanish oil.

McGovern operates in a tough environment. Profit margins are narrow, and major players dominate the scene. The food-service industry is supplied largely by a half dozen companies which get progressively bigger by absorbing smaller competitors. Sysco, the biggest, approaches $9 billion a year in overall sales. Supermarket distributors deal in huge volumes, and price usually matters more than subtle taste factors.

This bodes ill for any large-scale newcomer, whether from Spain, Greece, or California. Or an Italian outsider, for that matter. Importers may not have to worry about a visit from old Joe Profaci's goons, as in the old days, but the market can be just as tight. Pressure is more subtle, and it is legal. For one thing, there are slotting allowances.

"If you want to get into Safeway," McGovern said, "you've got to pay rent. All the big chains demand it, unless you're a very big guy. They figure that if you're a little guy with few items, your stuff is going to sit on the shelf. This can cost you 2 percent, and the margins are razor-thin."

Pricing is a nightmare for a product that is so dependent on nature's whims. Even if there is a bumper crop, olive oil cannot be stocked in reserve for long. The Mediterranean is a small sea. If winters are too harsh, if the rains are badly timed or, worse, don't come at all, prices skyrocket. When God is in his heaven and the trees bloom in unison, prices plummet.

During 1994 the Spanish crop was awful, and in 1995 it was worse. Italy did well in 1995, but the Greek crop was so-so. Oil production fell by 30 percent one year and went down another 50 percent the next. A gallon that cost wholesalers $8.50 in 1993 cost nearly twice that late in 1995, and the price was climbing by the week. By 1996, McGovern was paying up to $20 a gallon. But rain in Spain would change the picture dramatically.

Whatever the circumstances, the Italians stay on top. "Italy manages to undercut everyone in the U.S. market," he said, not without some admiration for smoke-and-mirror tactics no one seemed able to explain. "It's sort of like Milo Minderbinder in *Catch-22*. He buys eggs at four cents each, sells them for three cents, and makes a profit."

Part of the explanation is cheating, McGovern said. Dishonest packers cut olive oil with cheaper seed oil. But this is hardly confined to Italians. American importers sometimes cut oil once it reaches the United States. "Especially when olive oil gets super-expensive, people do a lot of slimy stuff," he said. "They try all sorts of creative ways to stretch good oil."

One example he mentioned involved a multimillion-dollar lawsuit in Los Angeles. Will Sauro of Tama Trading, a third-generation olive-oil dealer, had sued four competitors, all American importers. Independent tests showed some had cut olive oil up to 90 percent with canola oil. Sauro's lowest price was $100 a case; theirs was $49. "We complained to the FDA," said Nancy Fitzhugh, Sauro's lawyer, "but they were comatose and unresponsive." FDA officials explained that they had higher priorities for their limited resources. Testing oil can cost $700 for each sample.

Richard Sullivan of the North American Olive Oil Association condemns adulteration but faults restaurateurs who should know what they are buying. He knew about the Tama Trading suit. "One of the defendants told me, 'I'm a crook, but not in a wrongful way,'" Customers, he explained, seemed happy to get something marked "olive oil"—whatever it was—at half the price.

Most Italian exporters are honest, McGovern said; their advantage is skill and experience. The successful ones have mastered all the steps: negotiating, shipping, blending, and selling. They blend inexpensive-quality oils and package them at low cost. "A lot of their mass-market oils don't taste half bad, and they manage to do it cheaper than anyone."

In the short term, he said, California growers have an obvious niche at the top end of the market in specialty stores. But nothing prevents them from expanding to substantial production at reasonable prices. You only have to look at the Sciabicas of Modesto, whose cottage industry is headed for bigger things.

Dan Sciabica loads his oil into a white van and drives to the Saturday-morning farmers' market in San Francisco, just as Jeannot Romana does in Draguignan. The parallels are striking. Not long before the Romanas left Italy to grow olives in Provence, Nicola Sciabica settled his family in rural

California, where they grew olives. Dan's father, Joe, is also past eighty, still hard at it.

Most striking is how Dan and Jeannot resemble each other: the same hazel eyes, curious and friendly, always quick to light up. Each works a long, hard, and dirty day without a thought to billable hours. Either one could be trusted with only a handshake. And they both love their olives.

Even their markets are similar. San Francisco's is spread out, open-air, on a parking lot by the old Ferry Building. Table after table is piled with organically grown vegetables—exotic greens, fat purple eggplants, lush tomatos, portobello mushrooms, chilies of every stripe—sold by the people who raised them. There is milk and honey, crabs and oysters, cheeses and chocolates. People line up early for fresh-baked bread. Someone is roasting Anaheim peppers in a big round drum. A noted chef is giving *haute cuisine* lessons at a tiny outside auditorium.

But backwater Europe is a different world from country-style California. Each week Jeannot fills a crate of round-shouldered glass bottles and corks them. He does not bother with a label. Most of his customers know the product. Others will buy or not, depending on the color of the oil or the mood they are in. He'll get the equivalent of eleven dollars a liter for oil that belongs up there with the world's best. At the market he chats with passing friends, pausing from time to time to make change and say, "*Bonne journée*," to a disappearing back.

Dan Sciabica takes longer to load his truck. He offers a half dozen varieties, each in classy etched-glass rectangular bottles. Hand-lettered signs offer some details, but he talks nonstop, all morning long. He tells people about cooking, but also about health. Most of his customers seem less interested in taste than in staying alive. Among the Sciabicas' most popular oils is something they call "liquid butter," a sweet, soupy fluid from missions picked in late May. For its purpose, it is terrific. At twenty dollars a liter, it is also cheaper than the others.

Shoppers must take Dan at his word. The San Francisco Health Department forbids olive-oil tasting in the open air. It also requires all the meat, fish, and cheese to be sheathed in plastic, giving the farmers' markets an air of Safeway. I bought an armful of oils to check the differences, figuring I'd unload it fast; olive oil—even opened—is a welcome gift.

The Sciabicas continued to press oil in Modesto when it was at the depth of fashion. Like the Romanas, they made a limited amount and sold it, bottle by bottle. Now, with new labels popping up all over Northern California, theirs is still the best. In fact, they supply free advice on olive growing to many of their new competitors and, for a fee, press most of their oil. And their market is growing fast.

"In the sixties, the seventies, we could only sell to a few Italian restaurants," Dan said. "No one else wanted olive oil. We went through some pretty hard times. Now . . ." He fanned his hands skyward and smiled. "Fifteen, sixteen years, he couldn't sell green oil. Now everyone wants it. We can't keep enough." In a good year, the Sciabicas make fifty thousand gallons.

At the market, some people listened attentively to Dan's patter and then moved on without so much as an acknowledging grunt. Others, however, happily forked over $25 dollars a liter. "A lot of those who walk away come back later," Dan said. "And once people switch to good oil, they tend to be regulars."

The house blend is called Marsala, named for the Sicilian seaport Nick Sciabica left behind, produced every year since 1936. It is all California, a mix of mission and manzanillo olives picked after the color turned. There was also a pleasant little picholine, a few shades sharper, to liven up a salad dressing.

I asked Dan about some bottles that looked out of place, of a different shape, with labels half in Arabic script. The brand name was Almazara, the Moorish word for olive mill now used in Spain. They had pressed the oil from someone else's pendolinos, which had been trucked up from Caborca, Mexico.

Now tell me olives are not universal: here were Tuscan olives grown in northern Mexico, labeled in Arabic with an Andalusian name, pressed in central California by Sicilians, and sold along the San Francisco Bay.

Dan handles marketing, and his brother, Nick, runs the mill at Modesto. But Joe is boss. For some reason I'd expected an enfeebled patriarch whose periodic kibitzing was tolerated by indulgent sons. Not exactly. He works more than a full day, bustling around at turbo speed. In corduroy Levi's and a rakish striped shirt, with a smooth tanned face and handsome white hair, he hardly looks eighty. "If you eat well, you live a long time," he told me. "I have an aunt in Modesto who is 103. She walks better than I do."

As I followed Joe around town, his back pocket rang repeatedly, and he rapped out instructions via mobile phone. As we hurried to the Ford dealer to pick up a repaired truck, he was distracted. His 1982 Pieralisi press had stopped, and he might need a pump from Italy. Also, a pair of olive-oil industry promoters from Rome were visiting.

"I'm kind of a diehard, not in it just for the money," he said. History is proof enough of that. When his neighbors cleaned up growing almost anything else, he stuck with olives. Now his advice is in heavy demand by neighbors who want to plant trees and cash in on the boom. "The truth is, we went through a long, dark tunnel. But I think we're into the light now. I'm glad I lived to see it."

Joe invited me in while he dazzled his other guests with olive talk in rapid-fire Italian. One after the other, we tasted his oils with bread baked over old grapevines. The Italians seemed genuinely amazed. "Someone wanted to know if we put apples in it," he said of one late-season blend. Another, freshly pressed, sat us up in our chairs. "It's better when it bites," he observed. The Italians took another taste.

"Listen, stay for lunch," Joe commanded. "Just a simple family meal." Joe made yet another phone call, and we hopped into our cars. Minutes later, Joe's wife, Gemma, laid out a fabulous meal for five of pasta, lamb, vegetables, and an oil-laced salad of yellow golden globe tomatoes. I asked her how she did it, and she laughed. "After fifty-two years married to that one," she said, "you get used to it." Gemma is the company taster. If she does not like a batch of oil, they don't pack it.

In the old days, work was nonstop. When his son Nick was born, Joe had to call his cousin to come down and run the separator so he could go to the hospital. Dan, who is seven years younger, grew up loving the work. He used to hang around with a chunk of bread for a raid on the first oil off the press.

"I feel that I'm selling something that is healthy for you, and that makes me feel good," Joe said. "That's why we're able to do what we're doing. I hope I can live a little longer and enjoy the fruits now that it is so invigorating."

Joe was born in Waterbury, Connecticut, not long after his father landed, the first of seven brothers and sisters to emigrate from Sicily. He has yet to make it to Italy. Nonetheless, their table might have been in Marsala, down to the homemade *biscotti*, *taralla*, and Sambucca with the back-on-your-feet coffee.

Nicola Sciabica moved onto the property in 1925, after four years in San Francisco, when Modesto had twenty thousand inhabitants. Since then, a half million people have moved in around them. The city took seven of their twenty-two acres for a park. Now their olives come mostly from a ranch they bought in the next county.

In his long life Joe has seen big changes in the business of growing olives. He does not like many of them. Even with the new popularity of oil, he said, official interference has made it far too difficult. I mentioned the excessive safety rules that plagued Reisinger at Olive Hill. Joe was steamed. "Freedom has gone in America," he said. "It was different when you had to work if you wanted to eat. Now there is all this free stuff. No one wants to work. Everyone is overprotected. With all the restrictions they got, people don't climb ladders. We should get a hundred tons from our trees, with Italians picking, but with all the rules and regulations, that's impossible. There's a limit on how close to the road you can work. All that kind of stuff."

Workmen's compensation ate deeply into the thin profit margin, he said. Insurance premiums were high; workers were likely to sue on the slightest pretext. Minimum-wage levels prevented growers from hiring kids part-time. Tough immigration policies were drying up the pool of seasonal migrant labor.

"If it weren't for the Mexicans, who'd pick the fruit? They're the only ones physically fit enough to do it."

There was nothing wrong with protecting workers, he said, but regulations too often missed the point. As a result, growers cut back or went out of business, taking with them the jobs from which the workers were meant to be protected.

"When government people step in, they don't know what they're doing," he said. "You can be the smartest person in school, but without practical experience, that's no good. What do they know about olives?"

After six decades of experience, Joe took over his father's *fiscoli* press—and his name; most people call him Nick, because the company is still called Nicola Sciabica & Sons. Expansion convinced him to buy a continuous system that mills five thousand pounds an hour. But he runs it below speed. "Otherwise," he said, "it's like eating too fast."

The Sciabicas are careful to keep the water in their press at room temperature. Earlier, one of the Italians had treated me to a condescending lecture, insisting that no one could make oil without hot water.

As we drove along, to make conversation, I mentioned that a few short bursts of hail had hammered my olive crop.

"You, too, huh?" he said. He had suffered from hailstorms almost at the same time, when the olives were too fragile to withstand them. Then we found reason in common to curse the cochineal and its dreaded black scale.

I left Joe Sciabica with two crises on his hands: His foreman could not repair the pump. Olives, washed and waiting, could not be pressed, and the season was just rolling into gear. Here he was in deepest California with the nearest spare part somewhere in the Italian boonies. Worse, he had mislaid his trademark cap, a high-fronted, floppy cloth job. "My wife made it for me," he explained, looking distraught.

The following night I saw him at a dinner in San Francisco, surrounded by a pack of admirers. He had found a pump that fit his press. He had also found the cap, but he was wearing another one of similar design, his dress model. Bright turquoise with a vest to match.

That dinner, at the Palio di Asti, was thrown for the local olivegarchy by the visiting Italians and the Northern California Olive Oil Council. Before the speeches and food, a sizable crowd mingled in the bar to talk shop.

I found the oil maker from a hot new restaurant in Mill Valley called Frantoio, which means oil press in Italian. The owner, Roberto Zecca, had installed old-style machinery behind a glass wall. Signor Pieralisi himself came over from Italy to supervise. A retired international banker, Zecca had made oil in Tuscany. He wanted to feed Californians with flair. Diners could eat well while watching fresh oil spill down from a stack of *fiscoli*.

I'd been to Frantoio and liked the food. It was early in the season; the only olives to press were plum-sized sevillanos, which produced a watery oil that stung with chlorophyll. But it was a great idea.

Paul Vossen, a well-liked olive expert from the University of California, was surrounded by an admiring cluster. He is one of the state's few technical specialists. Vossen estimates that 34,000 acres in California are now planted in olives. He had mapped out forgotten groves, some masked by century-old oaks. "It is exciting to see these olives come back," he said. "But so much practical knowledge has been lost. People are having to relearn everything."

I looked around until I spotted Darrell Corti, whose family has run a specialty market in Sacramento since 1947. California olive people speak of him as they might a cardinal. The last person who mentioned Corti to me did it with an intriguingly goofy grin. Another said, "He's very nice, but you have to know him." Still another volunteered, "He knows a lot. Not as much as he thinks he does, however."

Corti was the one in a serious suit, with a sedate silk tie. He smiled pleasantly, with that lovable European trait of assessing his interlocutor, just obviously enough, down a slightly angled nose. His Italian was fluent. He had mastered Old World cultures and perfected an attitude of self-confidence in his opinions that the ungenerous might call pompous.

At first he brushed aside California oils, noting that the Italians had had a little more practice. Then he said that all California oils were good. I was confused.

Hoping to head off unnecessary fencing, I recalled for him my Spanish friend's answer to the question of what is the best oil: "What is the best cheese?" That said, I asked him what he thought were the best oils. It depends on your taste, he replied. I tried a different tack.

"Okay, I walk into Corti's and tell you I'm looking for a good California oil. What do you advise?"

"I'd ask what you wanted to do with it."

"I'd say maybe salads, general cooking, but I really wouldn't know."

"Then the question would be, why do you want a California oil?"

"Okay, okay," I said, beginning to sound like a Joe Pesci character. "Personally, what are your three favorite oils?"

He named Italian estate brands from Tuscany, Umbria, and the

Marche, all difficult to find without a seat on Alitalia. Then I tried a process of elimination. Did he stock Núñez de Prado, for instance?

"No," Corti answered, with more emphasis than I would have expected. He did not like the taste of the Spanish picual.

"Why not?" I asked.

"Because it has the aroma of cat's piss."

Fair enough. All the more for the rest of us. He left to prepare for his dinner speech before I got to Kalamata.

Wine snobbery is a fine institution. Why not oil snobbery? It is a complex business. Later, when I described the first part of my encounter to Nancy Harmon Jenkins, she laughed hard. She knows Corti. But then I asked what she thought about the picual. "I'm not wild about it," she said. "It has that whiff of sauvignon blanc." This is politely described as grassy, she said, but it means grass in which incontinent cats have frolicked.

Intrigued by Dan Sciabica's mention of pendolinos in Caborca, Mexico, I called Lawrence Johnson and went down to see his field of dreams. I had seen olives in every sort of setting, but I was definitely not ready for old groves against a panorama of tall saguaro cacti. Johnson was a California produce broker until the Tassajara Bakery needed someone to market their bread. The Zen Mountain Center in Tassajara Valley near Monterey was the first Zen Buddhist monastery in the Americas, but most people knew it for its ovens. The monks made every sort of bread, and visitors drove for hours to taste it. Johnson packaged it and put it in stores. By the time he moved on, the bakery had gone from a cottage industry in the trees to a thriving Bay Area chain.

Looking for something new, Johnson found olives. He first looked around Tubac, in the hills by the southern Arizona border, but gave up the idea. Too cold. He knew the string of missions that brought trees to California began in Sonora, and he dug out the old research. In 1909 two agronomists from the University of Arizona had crisscrossed the northern desert. They logged each stand of olives and made careful, often witty notes on the oil they tasted. This seemed like promising territory.

Then Johnson chanced upon Pancho Baranzini, the Mexican grandson of an immigrant from Lago Maggiore in northern Italy. Baranzini grew table grapes and almost everything else. He had some pendolinos and man-zanillos. Farmers nearby had a lot more. With money from a Tucson in-vestor, Johnson and his Baranzini were ready to go.

This, to me, was a noble project. Northern Sonora was my first love when, growing up in Tucson, I first discovered life beyond the U.S. borders. Its deep-fried tortillas and beans could be a perfect staple if they did not leave pockmarks on the digestive tract. But Arlene Wanderman, in New York, pointed out what should have been obvious. Before Mexican cooks found all that week-old grease, they used olive oil. Why not again?

I drove south from Tucson to the border at Nogales, on down sixty miles to Santa Ana, and then west across a saguaro-studded moonscape to the old mission town of Caborca. From there, Johnson took me over pitted narrow roads for a close look. We made a quick stop at the new plant, which housed an Italian continuous system and stainless-steel tanks, all gleaming from careful cleaning. We pushed on to Baranzini's ranch to see the old press. There we stayed.

Side by side, the two partners look like something out of an unlikely buddy film. They are of equal height, square-shouldered and fit. Johnson is clean-cut, fresh-faced, and fair, at home in well-cut slacks behind the wheel of his fancy Infiniti. Baranzini, bronzed beyond his skin tones, wears a black mustache and a straw cowboy hat, a knife sheath on his silver-buckled snakeskin belt. He bounces around in a dusty pickup. Each has a gentle smile, and they are very good friends.

Pancho's younger brother, Mario, runs the old press, which turns out only a small portion of the company's oil. Together, the Baranzini brothers work with farmers in the region who bring in the pendolinos, nevadillos, missions, and manzanillos (manzanitos down there) and pajereros they buy by the kilo. Quality control is their greatest problem. Anyone's bad olives can ruin a whole batch.

Their small Italian press is right at home among the cactus-rib corrals and crumbling adobe. It might as well be driven by burro. Olives are crushed in a small, battered machine using the old-style hammer method, perhaps the most efficient of all. Mash is spread a little sloppily onto nine *fiscoli*. When the short tower is locked in place, the old press soldiers on until oil flows slowly down the mats.

The decantation process is along the lines of the one used at Ekron, but the Philistines did it under a roof. Oily water goes into one fifty-gallon drum. After half a day, it is drained from the bottom, and oilier water is transferred into another drum. Cheesecloth strains out loose bits. Finally, a lanky Mexican in low-slung Levi's hops onto a forklift and hoists the last barrel onto a rickety wooden apparatus, like a camp shower. Pure oil runs down into a tank via a makeshift spillway of pipes and troughs to a glass filter.

Workers pour the finished oil into bottles and seal the corks with foil wrapping that was heated in an old saucepan. They pack some of the oil in five-liter tins. At full bore, the process can turn out three tons in twenty-four hours. From fall to January, they might total three hundred tons.

This was the Mexico I knew and loved, funky enough to send efficiency engineers and health inspectors into fits. But it worked. The oil ended up perfectly clean. Baranzini's variations on a theme respected the same principles his forebears used in northern Italy. And this was much more fun to watch.

Then we gathered in a small adobe-walled room for tasting, an old Mexican version of the Núñez de Prado breakfast ceremony. Pancho warmed corn tortillas over a gas burner exactly the way Paco in Baena had heated his bread. Instead of capers, we had a bowl of giant seedless raisins, the pride of the Baranzinis. Small glasses of combustible fluids were poured from unlabeled bottles. We chatted and joked in singsong Sonoran Spanish.

The oil was like nothing I had ever tasted. Fresh from the press, it burned like jalapeño juice but with a full herbal flavor; it would mellow nicely over the months. Pancho studied my face, pleased when he saw their visitor liked it. For an hour he asked detailed questions about how the outside world made oil.

Like so many growers, Johnson has high hopes for the future. He is counting on an end to European Union subsidies, which would make American oils far more competitive. He is convinced that once people taste his oil and word of mouth begins to spread, he cannot miss.

"We want to come out at a fighting price," he said. That would be $5.99, or even $4.99, a half liter. Beyond paying off the new press, his costs were minimal. Labor was cheap. The trees were there, protected from olive fly and other pests by the hard sun. The bottles and labels were elegant but simple. The market was only a truck ride away.

There is competition from the Phoenician Olive Oil Company, which presses olives grown on the Gila River Indian Reservation between Tucson and Phoenix. It was already on sale in Arizona in fancy amphora-shaped bottles. But Johnson worked out a marketing *modus vivendi* which would help both of them flourish.

The only problem was the olives.

Most Caborca farmers have refined the concept of *mañana*: their version is like *mañana* elsewhere in Mexico, but without that sense of urgency. Also, at least one generation has passed before any has gathered olives for profits. Neither tradition nor pride is at stake. Farmers might wait until olives fall by themselves and then scoop them up, dirt and all.

Other growers might try hard to pick their trees thoroughly but cannot reach most of the olives without a helicopter. Few pickers have the skill to harvest with a stick. Most trees are too bushy to climb. A granddaddy mission tree might produce 150 kilos of olives in a good season, but its owner would be lucky to collect thirty.

In the dry desert hills around Caborca there is no surface water. It rains little, and then usually in the fall, when it's too late for olives. The mere presence of groves is a miracle even southern Tunisians would be hard put to match. Yet most of the trees are drowning.

To help grow impossibly thirsty vegetables, the government subsidizes the pumping of ancient fossil aquifers far below ground. Farmers use the nearly free water to flood their trees, trapping it in dikes of heaped-up earth. Olives grow fat, but their oil content is seldom more than 5 percent. In field after field I saw tall old trees in calamitous shape. No one bothered, or knew enough, to cut away suckers. You could hardly see the main trunks. Mostly, the trees were choked with excess foliage.

On some ranches, however, old groves went unwatered and untended, left at the mercy of hot winds and the weeds that stole their nutrients. One looked as if Pancho Villa had ridden through with a torch, leaving burned-out ruin behind.

Finally, we stopped at a last grove, the property of an important farmer. Trees had been planted in neat rows far back before anyone could remember. Neat pump-fed irrigation ditches ran among them, banked by earth bunds.

The trees were bad beyond belief. Decades of neglect had left them looking like banyans after a high wind. Each tree had a half dozen trunks rising off the same tired root. From the ground up, bushy tangles spread every which way from heavy misshapen limbs. Many branches were dead of disease or drowning. High up in one giant tree a supporting bough had simply snapped off from the weight of unpruned foliage. The shattered, exposed ends were so waterlogged they looked as pithy as balsawood. Knee-deep water sat in puddles around the trees, going nowhere because the ground was too soaked to absorb it.

"We will need a lot of work with the farmers," Johnson said ruefully. "A lot of work."

It was November when I met Lawrence Johnson, nearly time to start picking back at Wild Olives. But my crop was a casualty of the Bosnian War. A sort of peace had been arranged, and I would be spending Christmas somewhere east of Sarajevo. Not that there would have been much to pick. Those two bursts of hailstones I mentioned to Joe Sciabica had knocked off half the olives on our hillside. Hard rain at bloom time did not help. But I'd had some fruitful moments on the road.

By the time I finished my research, I had gone to a dozen olive-growing countries and had studied a dozen more. In the end, however, my last few steps on the olive trail led me home again. Not to southern France, but home. Along Park Avenue in Tucson, just inside the low stone wall bordering the oldest part of the University of Arizona campus, I walked under a long row of magnificent trees. From their size it was clear they had been there since what passes for antiquity in America, at least since Wyatt Earp rode to Tucson to finish up the gunfight at the OK Corral. During my four years at school, I had barely noticed those trees. Back then, I could identify a saguaro cactus without much trouble and probably a eucalyptus. Olives were only something that went squish underfoot on my way toward a cold can of Coors on a late-fall afternoon.

This time I looked more closely. The soaring trees were pruned for sweep and beauty. Generations of expert hands had shaped solid scaffolds, faithfully whacking away low growth until the trees got the point. Huge roots sent vitality straight to the top, wasting no effort on useless suckers. Gnarled trunks twisted and bulged, just as they were meant to. Dead wood was sawn away, leaving rich green foliage to shade passersby. I felt that same air of peace—breezes through the olive branches—that I'd come to love around the Mediterranean.

It was, of course, Tucson. Overweening town fathers had banned new olive trees within city limits, alleging that their pollen aggravated people who had sent their sinuses to Arizona. But in the heart of town, and behind the older homes that climbed the foothills toward the Catalina Mountains to the north, fine trees produced their yearly crops. And these olives on my old campus were among the finest I had seen anywhere.

Looking eastward at the weathered red brick buildings, I could not miss the irony. Humanities, back then, was forced servitude. Two hundred freshmen and I sat gazing out at the afternoon sun while some teacher droned on about past lives we could not picture. The Peloponnese, Asia Minor, the Something or Other Maximus were names to be committed to memory until final exams. But they were all just out the window the whole time. Had we walked outside and sat for a while under an olive tree, we might have understood most of what we needed to know about the roots of Western civilization. It worked for Plato.

Cailletier (Niçoise)

$$\overline{16}$$

Another Season

Our whole hillside missed the Christmas picking. It rained, very nearly forty days and forty nights, and no one went out to their trees. Ancient rock walls collapsed. Mud slips carried away whole trees, roots and all. And then the sodden ground began to settle. The narrow lane we all loved collapsed into one of those underground water sources that Pagnol's Jean de Florette tried so hard to find.

One Saturday an old man we all knew only as Jacques drove out to check on the magnificent old trees he tended with care. They were just below the road, so beautiful that passing drivers often stopped for a look. Three had dropped out of sight into a sinkhole, without a leaf visible aboveground.

I was off working. By the time I straggled back to Wild Olives, 1996 was two weeks old. Half my crop had been lost during that August hailstorm. The rest of it was lying on the ground.

Strangely warm weather had tricked my wily old trees. Emiliano, sniffing spring, sprouted light-green shoots from his freshest wood. Ernesto, deep in a pool of standing water, did not know what to make of things. This had the makings of catastrophe. A sudden hard freeze could throw the neighborhood into mourning.

That is how it is with olives. Natural rhythm plays its part, but so does the chaos theory. Far to the west, Jaén farmers grieved over a drought-crippled crop that was a quarter of what it should have been. At the same time, they rejoiced. Those same rains which had washed us out had saved their groves. Next year, the Virgin Mary willing, they would have a bumper crop. Across the Mediterranean, on Jerba, Mhenni Ben Maad's harvest was mediocre. But Allah would intercede. In the wild Mani in Greece, trees were full. Because of everyone's hard luck, farmers raked in the profits. Next year, it might be the other way around.

Down at the Flayosquet mill, Max Doleatto scooped decanted oil *à la feuille* with his habitual half smile. "Two-thirds of the olives are on the ground this year," he said with a shrug. "But the ones we have are terrific. Taste this." A barrel of fresh caillet roux oil was nearly gone, dispensed in five-liter containers to the usual suspects. I bought mine and paid fourteen dollars a liter for it. Max would make it just fine.

With Fabrice Godet at the Draguignan mill, it was the same. No one had many olives, *hélas*. Then he changed the subject. He wanted to go the Croatian islands off Split in March. Did I have any suggestions? In Aups, Charles Gervasoni just sighed when I asked about the season. They knew too much about olives to be surprised, or depressed.

One January morning the sky came up cobalt blue, cloudless, and the clear air had the snap of early spring. Jeannette and I found Paul Bosquet's crew at work in his biblical orchard up the road: his wife, Marie Romana; her brother-in-law, Roger Martin; their friends the Fabres, who ran the Cotignac mill; and Fernand Fabre's cousin Yvolle, who had learned to walk and collect olives in the same year. We scrapped our day's plan and joined in.

By noon it was nearly shirt-sleeve weather. Sliding those fat red caillet roux olives into baskets was like picking cherries in May. We did the big trees the old way. As soon as large black nets were spread underneath, a dozen flashing hands dropped olives onto them at a furious pace. Fernand combed the branches with his little plastic hand rake. The rest of us preferred to feel the olives and leaves.

Yvolle told stories in incomprehensible Provençal. One was plain enough: she was there for the 1956 freeze, and she had wept at the sight. The Bosquets and Fabres, laughing loudly, relived their recent trip to New York and Niagara Falls. America was wonderful, they allowed, but they got homesick at dinnertime.

I climbed high into a beautiful tree, which rose like the pedestal of some great sculpture, from a massive base into three soaring limbs that supported a willowlike crown. For a moment I basked immodestly in Paul's praise: "Ah, the American champion." Then I heard a rustle and looked up to see Roger's boots over my head, along with a thick shower of red olives.

We used a long ladder, a single pole with holes drilled for footpegs that extended from either side. "It has to be white pine," Roger explained, "and the cross pegs are cut from a thick genêt trunk. But you have to do it in the right moon. When it is dark. Otherwise, *les bêtes* will eat it to pieces before the first season."

He did not know why the moon phase mattered, he just knew it did. A little insect poison would also do the trick, but then a store-bought aluminum ladder probably would, too. That was not the point. Years ago, no one had the choice. Now these people don't want it. They were not out there because they could not afford cooking oil or bottled salad dressing.

Occasionally Roger would marvel at a tree, as if he had not seen enough of them over sixty years. One old monster trunk had frozen on one side, and the centuries had stripped much of it to bare wood. But a narrow, living vein still carried sap to the top to support thick foliage. "It is all in the *peau*," he said. Skin, not bark, as though he were talking about something human. "As long as there is a little bit that's healthy, that's enough."

Then it was lunchtime. Marie spread burlap olive sacks on the clover under the trees. On top, she laid out a gleaming white linen tablecloth. Paul built a fire in his shepherd's Weber: three rocks supporting a blackened grill. We started with home-cured ham and pâté as the logs burned down. Next, Paul toasted bread on the olive-wood coals. Marie rubbed the bread with garlic and dribbled on figure eights of olive oil flavored with anchovies.

We finished with wild pork chops, carved from a *sanglier* someone had shot down the hill a few days earlier. Just in case, there was also lamb. The Bosquets had not counted on drop-in mouths to feed, but there was plenty left over. Paul poured nonstop from unmarked liters of red wine that I would not have traded for Rothschild's best. And small corked bottles contained the season's first oil, *huile trouble*, with a harsh bite that French oils are reputed not to have.

After homemade cakes, tangerines, and bolt-bending coffee heated over the fire, we contemplated the afternoon's picking still ahead. This year, it was a party. When it was cold as hell and sleeting, it was penance. No one thought of who owed whom, or counted the hours. It does not work that way.

Somehow it is always worth it. That year we had no olives on our trees. But with the grace of God, or those Egyptian and Greek and Roman deities who have protected olive lovers forever, oil from those sacks of caillet roux would be the ingredient of honor at the next Fête des Jumeax in June, this time at Wild Olives. Emiliano would impress a few of our family on the hillside, and they would be too polite to snicker at his pals.

Roger Martin must have read my thoughts. *"Vous savez, l'olivier, il fait ce que il veut."* The olive tree does what it wants. He glanced around and smiled. *"Mais, c'est beau."* He meant all of it: the great old living monuments, the stone ruin nearby where his family once lived, the gorgeous winter-spring day, country friendship, and the remarkable little drupe to which he had devoted a very full life.

Yes, I thought. *C'est beau.*